# Nuclear Mustang

# Leadership

By

WT Subalusky

# ABOUT THE AUTHOR

Career crew member on several U.S. naval submarines, executive in a highly technical industry, coach and mentor of supervisors and managers during three different careers, the author has performed successfully in positions at all levels - entry-level employee, supervisor, manager, and executive. After enlisting in the U.S. Navy directly following high school and volunteering for submarine service, he rose through most of the enlisted ranks, received a Commission as a Naval Officer of the line while also acquiring both Bachelor's and Master's degrees, specializing in electrical, acoustical, and nuclear engineering, as well as completing advanced management studies at the Harvard Business School. Following his military career, he was employed in what was essentially a consulting role for a newly formed company that assisted the commercial nuclear power industry in moving from a historically poor level of performance that culminated in the Three Mile Island nuclear accident to a level of excellence unsurpassed in the world, where it remains today. In the course of this career, he observed and critiqued workers, supervisors, managers, and executives as he evaluated their performance against standards of excellence. Following this second career, he started his own consulting company and spent the next fifteen years assessing the performance of executives, managers, and supervisors in a variety of high-risk industries, including commercial nuclear plants, uranium enrichment facilities, new technology fossil power plants, and major hospitals. He also served on and chaired the safety and performance oversight boards of several major nuclear power companies in the U.S. and Canada and authored the books *The Observant Eye - Using It to Understand and Improve Performance and The War on Error.*

# Table of Contents

# 2

# 3

# PROLOGUE

One of the most interesting things about life is what can be seen when one looks, and I mean really looks. By "look," I mean a concentrated observation, with a critical eye, that identifies whatever can be done to improve performance in the activity or condition being observed. I have written a book on this concept of critical observation. The book's title is The Observant Eye, and it includes examples of some of my observations conducted over several careers in the U.S. Submarine Service, commercial nuclear power, and other high-risk industries, a few of which will be further discussed herein. If the Observant Eye is part one of how to run a better organization of any type, this book is part two. Part one is about finding problems; part two is about fixing them and preventing others. This is a book about fixing and/or preventing performance problems through leadership, a kind of leadership that, unfortunately, is not seen often enough.

I chose to start here with a comment on observation because if one observes properly, or really looks, to use my words, into just about any activity or condition, there is a likelihood that shortfalls will be seen in the kind of leadership this book is about, Nuclear Mustang Leadership. This kind of leadership will be described further in a bit.

Hopefully, this book will not only be useful to leaders and would-be leaders but will be interesting as well, with the interest piqued by the fact that it is replete with true stories and happenings drawn from the deep well of my personal experience. Through these stories and tales, I share advice regarding leadership. They include what I saw when I really looked and will hopefully both motivate and assist the reader in achieving the level and type of leadership needed to prevent and/or correct the issues seen.

I'll define Nuclear Mustang Leadership in a bit but first, one of my stories - told in the words any worker might use.

I was watching some work activity that involved little more than installing a small part by fastening it in place with a screw. Where this work was being performed is the most important part of the story, and I'll share that shortly. The work was being done by a middle-aged gentleman who was, as evident by the directions he periodically shot out to no one in particular, the job leader. This was a small part of a larger and more complex evolution and the job leader was assisted by a similarly aged woman. The leader and assistant were being helped by four other people who provided various types of support, such as going for tools, making sure the work site remained relatively uncluttered, and performing other supportive activities. Before this evolution was over, the discussion between the leader and the assistant would, unexplainably and in the presence of the others, gravitate toward a very candid discussion of personal sex preferences. Where this all occurred might be even more surprising than the fact that it occurred at all.

As the job leader began drilling the hole for the subsequent installation of the screw, problems began to develop. Remarks by those involved indicated the drill bit was not sufficiently sharp. As a result, and because of reasons not apparent, it was decided by the work team not to get another, perhaps sharper, bit. As a result, the relatively dull rotating drill bit then had to be forced and repeatedly tilted in one direction and then another to fully penetrate the material being drilled. The hole was finally completed to the satisfaction of the job leader, and the screw that was to hold the part in place was prepared for insertion into the hole. The roughly drilled hole had not been adequately prepared, and in short order, the screw got stuck. Once again, force was applied, this time to the screw. The application of force continued until the screw became even more firmly stuck. The fastener was only part way into the hole and could not be moved, either in the forward or reverse direction. Efforts to drive it either into or out of the hole ended when the slot in the head of the screw became deformed and was no longer functional. At this point, the job leader directed someone in the group to go and find what I would call an "easy out." This is a device that can dig into the head of a stuck screw and allow torque to be applied such that a stubborn and damaged fastener can be rotated and thus retrieved. During the entire course of this evolution, the job leader and his female assistant were carrying on a casual conversation unrelated to the work at hand. The science of human error reduction has shown time and again that focusing on the task at hand is an important element in reducing the likelihood of errors. This conversation displayed ignorance of this basic error reduction concept. More notable was the topic of the unrelated conversation. The pair was trading thoughts and personal views of oral sex. The coarseness of the discussion topic was considerable, and there likely was at least some embarrassment among the subordinate surrounding group, which contained both males and females, who were not engaged in the discussion but obviously could hear it. Any discomfort would have peaked when the female assistant loudly announced her position on the topic by stating, "I believe it is better to give than to receive." Unfortunately, the announcement was not questioned, and as the bawdy discussion continued, so did the search for the easy-out. After about fifteen minutes and the unsuccessful completion of two searches for an easy-out, each one ending with the designated runner returning to the work area only to report her failure to find the desired tool, the search was canceled de facto when the job leader chose to move on without the needed tool. In evident frustration, the job leader once again began dealing with the stuck screw using force - even more force than before, this time applying the force of impact with a hammer. He banged and banged on the screw with a force sufficiently great to cause tables and cabinets in the area to shake and shudder. (A person would hope that work of this quality would not be performed on one's car in the absence of owner oversight.) Although the brutal attempt to resolve the issue took much longer than originally planned, it was eventually successful in that the screw inserted enough to do its job at least in the opinion of the job leader. It was at this point in the evolution that the patient began to awake as the effects of her anesthesia slowly dissipated.

The above activity was a knee operation being performed in one of the operating rooms of a major hospital on a young lady in her early twenties. I was on the sidelines of the work, observing what was happening, working at the request of the hospital CEO to better understand how the performance of the subject hospital could be improved. The unnecessary delays, caused by, among other things, poor work planning, lack of contingency planning, and poor maintenance and availability of surgical tools, resulted in the operation taking longer time than expected. Murmurs from those involved indicated that the patient could not be safely returned to sedation, and thus, the operation, toward its end, had to be "hurried up," to use the workers' words, in order to be completed. The patient might know nothing of what had transpired and, unless some other unforeseen complication arose, would return to her active life with nothing more than possibly a longer recovery time and an unusually sore knee, sorer than she would have expected based on discussions with others who had similar operations. The operative word in the previous sentence, however, is "might". There is a belief among some in the medical profession that what is called intraoperative awareness can occur. If it does, the patient under anesthesia will not be able to respond to commands but may recall some of what occurred during the anesthesia. For this reason, some medical literature suggests that because the medical staff may not know if a person is unconscious or not, it has been suggested that the staff maintain the professional conduct that would be appropriate for a conscious patient. What just transpired in this scenario, and what I had personally observed, could be categorized in many ways. I would describe it as a lack of leadership on the part of many. Think for a moment how you would have reacted if you had been on the sidelines watching as the above scenario unfolded –the lack of adequate preparation apparent in the missing instruments, the poorly maintained equipment that resulted in a hospital drill bit duller than any you would use on a home project; the distracting sexual banter back and forth while delicate, attention requiring work was being performed. And suppose that young patient lying on the table was your daughter? Think how you would feel and what you would have done. You are now experiencing the thinking of a leader steeped in Nuclear Mustang Leadership. Such a leader would not have let this happen.

I provide this example in the prologue primarily because it makes a point: anything that happens in the presence of a leader and to which that leader does not react is fairly considered by witnesses to meet that leader's expectations. But to be honest, I also share this story because it has key elements that generate attention - sex and violence. Later, I will bring into my anecdotal examples other tales involving such diverse but interesting items as hunks of human flesh and submarines about to be crushed beneath the sea. I provide these examples to make important points but also to maintain the reader's interest in a subject that, like most management treatises, has the potential to be otherwise a bit boring. My objective here is to share my thoughts on leadership with the reader and do so interestingly, perhaps even in an entertaining way at times. I will provide clear, concise, and practical suggestions on how you can excel in the art of interacting with people in a way that can help an organization achieve the highest levels of performance. These will be fresh perspectives, many of which you have not heard before, such as how to get people to do things you want them to do without needing to tell them or even to ask them; how, when you are responsible for people and are convinced that you don't have time to do the things you believe you need to do when you have no time left, you can *make* time. I will tell and hopefully convince you that there could well be a time when both you and your company will benefit from insisting on the equivalent of requiring someone to walk around with a silly object on their head.

There is nothing theoretical in this book. Everything you read here will have come from and be described as specific, real-life examples of principles tested over time by me, gained through the experience of multiple decades of interacting with hundreds of companies and organizations, encompassing fields such as submarine operations, commercial nuclear plant management, government technical facility operation, and healthcare in hospitals.

While most management and leadership texts find themselves as dry as the lunar surface, this book attempts to be different. It serves an entrée of important yet sometimes unexciting leadership principles but embeds them in a mentally tasty, interesting, and sometimes even exciting assortment of side-dish anecdotes and examples that combine with the entrée to provide one of the more mentally delicious meals that ever lay on the mental tables of leadership-oriented minds. At times you will think this book is a novel. It is, however, not fiction. Rather, it is a collection of real-life experiences woven into a fabric of advice, a map, a set of directions, a reference manual, a "how-to," a written guide for achieving improved performance through Nuclear Mustang Leadership.

A recent survey found there are more than 15,000 books on leadership in print. Knowing this, some might say that writing another one is unnecessary and a waste of a writer's time. I, on the other hand, would say the opposite, and this I would say for three reasons: First of all, the fact that there are 15,000 books already out there on a topic means there is a considerable thirst for information on the topic, for advice, for new thoughts, for topic-related experiences. Just as important is the second reason writing the 15,001$^{st}$ book would be a challenge, and the challenge is one of those things the kind of leader about to be described herein looks for. Finally, the third reason and the reader will have to be trusting on this one, this book is different. So many of the leadership books available, including so many that are well worth reading, deal with softer aspects of the art (and it is an art) of getting people to do things you want them to do. They so often deal with topics like relationships, vulnerabilities, or flexibilities. This book, in contrast, is directed at results improving performance and correcting problems. It is story-laden, and even more importantly, it is also directed at leading others, but also someone different from the target of those other 15,000 books yourself. Yes, I said, and I meant, this book addresses leading yourself as well as others to success, to achieving results.

Within this book is a multi-element formula for a proven form of effective leadership, a formula, not unlike a chemical formula, that combines essential elements of life in submarines, those same kinds of elements important to the high-risk work of properly operating commercial nuclear power plants, and finally, the perspective of something with a name indicating it might be a horse. However, be warned. This guidance is not for those unwilling to put out what some might describe as a lot of, maybe even an unbelievable amount of, effort. If you choose to hang in there and pursue this development, you have already met one of the criteria for the type of leader described in this book because you will soon find that the willingness to put out such an effort is one of the key differentiating characteristics of the kind of leadership discussed here. As has often been said, most good things in life, including the fruits of the leadership formula we're about to describe, don't come easily, but they are worth working for.

One element contained in the following paragraphs is an assortment of quotes related to the topic being covered at the time the quote is provided. I have included quotes because they are succinct in capturing some of the most important thoughts on a topic. I provide them mainly for that very reason, but just as importantly, these quotes and the thoughts they convey come from a wide variety of people, including sometimes myself, so they are also a means of showing that the issues one deals with in leadership positions, first, have been dealt with before, and second, are not unique to any one business or area. The wisdom conveyed applies to just about everyone in a leadership position, including those just trying to lead themselves to success.

A final element contained herein, and one that is essentially the most substantive element of this book, consists of some principles that are universal and have stood the test of use over time. They have been invaluable in allowing me to advance from being in one of the lowest levels in society to being a senior military officer, a senior business executive, and a highly successful and well-compensated senior management consultant. These principles have been tested in real-life experience, not founded on theories.

I first encountered these principles during my time in the U.S. Naval Submarine Force, and then again experienced, as well as applied them myself, as I worked in the commercial nuclear power industry and at the prestigious Institute of

Nuclear Power Operations. These were the principles that helped the commercial nuclear power industry move from levels of performance that led to the infamous Three Mile Island nuclear plant accident to the high levels of performance that continue today. These same principles, I subsequently applied in my nuclear power and other high-risk work consulting business as well as in many aspects of my personal life. This broad application convinced me that these principles can be applied to just about any aspect of one's personal as well as professional life. They work whether one is operating a submarine, managing a business, or leading people in any circumstance, but they work equally well when one is dealing with one's family, social issues of the day, or other significant scenarios.

Let's begin our journey toward understanding the special kind of leadership this book is about by talking about one component of that formula mentioned above, with you coming along on a ride aboard a U.S. Navy diesel engine-driven submarine. Although this type of submarine is no longer in operation in the U.S. Navy, being long replaced by the considerably more complex nuclear-powered submarines, the tale about to unfold happened on one of these submarines while it was operating. Only this time, you will be there. During your ride, I will relate a relevant-to-leadership story about something that happened on one of these undersea vessels, a happening that, unfortunately, is not that unusual on this type of craft and that is highly relevant to a component of that special kind of leadership. To better understand my "not that unusual on this type of craft" comment, which refers to the perils of submarining, know that Prime Minister Churchill once said, "Of all the branches of men (and ladies these days) in the forces, there is none which shows more devotion and faces grimmer perils than the submariners." Surely, there must be something we can learn from such people. But what is meant by "the perils"?

Before 1914, there were 68 submarine accidents, including 23 collisions with the loss of more than 280 crew members, 19 internal explosions, and 13 sinkings due to hull access points not being closed when they should have been. Following that period, during World War Two, 52 U.S. submarines were sunk, and 3,500 men died, with most never being found. This unfortunate loss tallied up to be a casualty rate of

20 percent, a higher casualty rate than any other military group during that period or any period since. For example, in the same time frame of that loss, the U.S. Marines, a military group for whom I could not have any greater respect and admiration, had a casualty rate closer to four percent. Continuing in time, between World War Two and the time of this story, there were eight more submarine losses or major accidents.

Evidence of the hazards of submarining continues even into the current time because the hazards of submarine service remain high, even during peacetime. Since the time this tale unfolded, the U.S. lost the nuclear-powered submarine, Thresher, with 129 crew members, and the similarly powered Scorpion, with 99 crew members. Also, during this time, the Russians / Soviets had at least four submarine disasters. In 1986, a Russian missile-carrying submarine had an event involving seawater entering a missile tube, leading to an explosion. This submarine was eventually lost. In another instance, in 2019, a fire aboard the Russian nuclear submarine Losharikl, operating north of the Arctic Circle where some of the boats I served on frequently operated, killed 14 sailors before it could be extinguished.

It is well known that serving on submarines is a dangerous occupation. Certainly, we can learn something about leadership from those who choose such work.

The following is a story that takes place in the year 1962. Nikita Khrushchev was the head of the Communist regime of the Soviet Union. Sirloin steak could be had for about a dollar a pound, and a quart of milk cost twenty-five cents. If any music were to be heard, it would likely be the crooning of popular singers of the time, like Elvis Presley, Ray Charles, and maybe Chubby Checker. Also, in the air at the time was the dark but melodic tune of a new folk song, *A Hard Rain's Gonna Fall*. But the gloomy thoughts of that song by the newly discovered Bob Dylan were not on the minds of some young men whose thoughts were focused on other things as they pursued their chosen vocation - serving their country underwater. Neither musical artists nor the cost of food was a concern to these young warriors, and no music filled the foul and marginally breathable air of the space they were now in. This story unfolds in the Stern Room or end compartment of a diesel engine-driven U.S. Navy submarine, operating on a mission somewhere beneath the sea, location classified. This room, or compartment as it was called on the sub, would normally have been the After-Torpedo Room; however, on this sub, the torpedo equipment had been removed to accommodate special intelligence collecting equipment. Much like many of the compartments on this submarine, this one also served as a space for some of the enlisted sailors to sleep.

Within this stuffy compartment was a young seaman sitting on one of the almost uncountable items of equipment crammed into this already small space. The boat, as submariners called their craft, was quietly cruising along, not unlike a shark looking for a meal, several hundred feet under the ocean surface. If one could have observed the sub from outside, it would have looked to be operating normally quietly, in keeping with its Silent Service moniker, but quite in contrast with what was happening in the innards of the vessel.

The young naval submariner, who had been sitting a few moments before, had sprung to his feet and now stood on the deck in the Stern Room. As with all diesel submarine compartments, this room was claustrophobically small, much smaller than those on the larger nuclear submarines of today, even considering the crew numbered only 78 as compared to the approximately 140 crew members needed to operate a nuclear sub. Pipes, handwheels, gages, electronic equipment, weapons, beds, or "racks" as they were called, and other gadgets were mounted everywhere, making the space seem even smaller. As with all diesel submarines, the compartment reeked of bad odors body odor, a slight whiff of sewage from the toilet facilities, yesterday's dinner blended with today's, the fumes of diesel oil, and several other unidentifiable odors. But the smells didn't bother those who operated this underwater war machine. It was just the way it was for those on diesel boats, and the smells, which permeated the uniforms of the underwater warriors, were often knowingly and proudly transported to shoreside establishments, particularly those most frequently visited since they dealt in the sale of alcoholic beverages. In these facilities, in particular the smells were worn as a badge of honor. They identified the sailors as submariners, and submariners were known to be special not only because of their unusual chosen vocation and the more rigorous requirements that differentiated the job from those in many other available fields but also because it paid at least a few more dollars in both hazardous duty pay and submarine pay. Although the boat sailors would never admit it, it was likely this compensation factor that made them more attractive to some of those females who hunted male companions, and even mates, in these same sailor-mooring areas.

In this story, the young submariner, now standing in the Stern Room with a look of alarm on his face, had been informally classified as a NUB since he first reported to the boat. He was currently still in that classification by the crew. NUB was an acronym for "non-useful body," someone not yet qualified in submarines and therefore unable to help in any meaningful way during casualties of this nature. NUBs are under continuous pressure from the crew to qualify in submarines, and are typically asked ten or more times a day, how are your quals (submarine qualifications) going? When are you going to qualify? This pressure is continued until either a person qualifies to wear the submarine dolphin pin, silver for enlisted, or gold for officers, or is deemed unacceptable and transferred off the boat and out of the submarine service. The so-designated NUB was surrounded by several comrades much calmer than he, which he found surprising, considering what was going on at the time. The Flooding Alarm, a rising and falling siren, had sounded several minutes earlier, actuated by a submariner in a different part of the boat who initiated the alarm by operating a small, red lever with a star-shaped handle. As with all of the key alarms, the unique shape of the handle of this alarm allowed it to be identified by feel. The concept of having alarms be identifiable by feel provides for an effective response should there be an additional casualty leading to loss of visibility at a time when an alarm needs to be sounded. The above flooding alarm had been quickly followed by the boat's announcing system blasting out the ominous but clearly articulated words, "Flooding in Maneuvering. Flooding in Maneuvering".

"Maneuvering" on this submarine was the compartment just forward of the Stern Room where the young sailor stood. It's important to point out that on most submarines, and this one in particular, all major compartments are on the same level and are laid out next to each other in a series, such that to get from one end of the boat to another, one must pass through every major compartment. There is no way around any compartment. Simultaneous with these audible and somewhat scary signals, as witnessed by the NUB, one of the young sailor's sub-mates had, without hesitation, jumped into the flooding compartment, quickly shut the hatch behind him, and dogged it down tightly. "Dogging" is a navy term meaning that, in this case, a large, two-winged handle had been rotated multiple times to cause several big, tooth-like projections, or "dogs" on the thick steel hatch, to clamp down and tighten it against the rubber-lined sealing surface. This dogged-down hatch would surely keep the water that was reported as flooding into that adjacent compartment from also flowing into the Stern Room.

The neophyte submariner was taken aback by the seemingly irrational action of his comrade, now trapped in the flooding compartment. He wondered what had made him perform what seemed to be a suicidal act - jumping into a compartment that was flooding and would soon be at a loss for any breathable air. At the same time, the observing sailor took personal and selfish solace in knowing that one of the only two emergency escape hatches on the sub was in this compartment, the other being in the forward-most compartment. Should the sub sink to the bottom due to the added weight of the flooded water, the newly minted submariner naively concluded that he had only to enter the escape chamber and go through the procedure for escaping a sunken submarine. This was a procedure on which he had been thoroughly trained and in which he had demonstrated proficiency using a large water tank at a submarine school in New London, Connecticut. The escape, he thought, would have him soon safely floating on the surface. In short order, he noticed that the entrance to the escape trunk had been blocked by a large steel plate. The plate was held in place by a strongback, a metal beam designed to hold the steel plate in place. The strongback, in turn, was bolted down by a series of large nuts, each about the diameter of a half grapefruit. Access to the escape trunk was completely blocked! And apparently, it had been blocked on purpose!

Discussions with his much calmer comrades, who had long ago qualified in submarines and were now focused on controlling the flooding rather than escaping, soon revealed that the steel cover blocking the only means of escape, was part of the submarine's rig for depth charge, a condition set because the sub was at sea on a mission. The steel plate was intended to prevent flooding should the escape hatch above it be blown open under the bone-jarring shock that would come with a depth charge sent down from above. As alarming as the flooding next door, at least to the young seaman, was the just realized fact that the wrench needed to remove the large nuts and steel cover blocking his escape was in a compartment forward of the one now flooding. The compartments being laid out in series meant that there was no way to get to the wrench without transiting the flooding compartment. Not only was the access to the escape hatch blocked, but it also couldn't be unblocked!

The frightened sailor was informed of the wrench's location by his colleagues who were clearly more focused on what was expected of them as laid out in the submarine flooding procedure than they were on the location of the wrench. The Maneuvering compartment continued to flood, and the rookie submariner, now the author of this book, continued to worry, not realizing that he was witnessing a key principle of Nuclear Mustang Leadership. But would he or others die before he had the first opportunity to benefit from this exposure? And what about the suicidal sailor who had purposely isolated himself in that flooding space? In time we will revisit this sailor's happenings and will learn that the behaviors described in this story are among those encompassed in Nuclear Mustang Leadership.

# THE SUBMARINER ETHOS

Because it played such a significant role in the formulation of my leadership thoughts and so significantly undergirds many of the points herein, I will provide a brief description of how, in my view, those in the submarine force think, operate, and lead.

Although it was not apparent while I was experiencing it, hindsight subsequently allowed me to see that the seeds of leadership fruit and its characteristics, for me, were planted in the fertile and nurturing environment of a heavily weaponed steel tube, often hundreds of feet below the ocean surface. As the seeds germinated, that tube moved clandestinely about foreign waters, carrying an orchestra of about 100 performers as they moved about with the precise actions needed to put out fires, stop flooding, prepare for collisions, recover from unplanned excursions to overly deep water, deliver special land warriors, collect intelligence, and, most importantly, launch weapons of war. This was life on a United States Navy submarine.

In 1919, in "The Story of Our Submarines," John Bower, a World War 1 submarine

Commanding Officer, captured well the essence of submarine life:

*"There is a Democracy of Things Real in the boats, which is a very fine kind of Democracy. Both enlisted men and officers in a submarine know that each man's life is held in the hands of any one of them, who by carelessness or ignorance may make their ship into a common coffin; all ranks live close together, and when the occasion arises go to their deaths in the same way. The Fear of Death is a great leveler, and in submarines, an officer or a man's competency for his job is the only real standard by which he is judged."*

Bower's words provide the framework upon which I believe the as-yet unpublished ethos of the submarine community is based, and the ethos is the set of beliefs a community lives by. Think about the following elements of this ethos as I see them and how they apply to your business, whatever that business may be.

- Trust others but verify whenever possible because anyone can make a mistake.
- Work as hard and as long as needed to get the job done.
- Show by your actions that you understand the goal is a mission accomplishment, whatever it takes.
- Train and practice until you feel that you can't possibly train and practice anymore, and then train and practice more.
- Never run away from a problem. Run toward it.
- Demand teamwork; it's essential for survival, truly, for submariners, a matter of life or death.
- Remember that if the boat dies, everyone dies. Act accordingly.
- Strive for excellence in everything. Tolerate nothing less than the best performance and the greatest effort. Room on a submarine is limited. There is none for shirkers.
- Be always ready to deploy, to take the fight to the enemy, to get the job done.
- Get used to sacrifice, it is a submariner's way of life. Remember, you volunteered.
- Nothing is so serious that at least some humor can't be found in it.

➢ Show recognition and pride to be an elite member of a special group of people with a remarkable heritage.

Mustang is a military slang term used in the United States Armed Forces to refer to a commissioned officer who began his or her career as an enlisted service member before becoming a commissioned officer. Having been a Mustang, I can assure you that one of the most important attributes of this category of person is the ability to relate to, understand, and appreciate the thinking and the efforts of those to whom the directions are given. Think about this point. It's not unique to a military environment. These attributes can exist in any organization. There are ways to achieve that understanding and appreciation for the workforce. This ability to relate to the workers is the essence of being Mustang-like, the most important attribute of the leadership style proposed herein.

Not unlike the mixing of chemicals, when habits like those encompassed in the submarine ethos are combined with those of a Mustang, the resulting chemical-like substance of leadership is one with power far greater than the sum of its ingredients, a type of leadership that can accomplish anything. And, again, these "habits" need not be in a military environment. They are just habits, and therein lies one of the characteristics of Nuclear Mustang Leadership, it can and does work in any environment. More words on the "Nuclear" element of this title in a bit.

Think about the above, ethos included, as it might apply to your business or even to your life. Is your workforce as dedicated to the mission that they willingly forego sleep and would give anything to accomplish it? When a problem arises, do you have to designate someone to fix it, or do people, without direction, run to it, take it on, and do whatever it takes to fix it? When someone does an important task, do other members of the team back her up just in case she makes a mistake? Do people, without being told, work as hard and as long as it takes to get the job done? And then when they do finish the job, do they move on and take on another job without waiting for some big congratulatory celebration? Or even wait for further direction? If not, maybe your organization needs some Nuclear Mustang Leadership.

So why add the term Nuclear to the words Mustang Leadership? Read on to find out.

# SUBMARINE LEADERSHIP

*"Submariners are a special brotherhood; either all come to the surface, or no one does. On a submarine, the phrase all for one and one for all is not just a slogan, but reality."*

-VADM Rudolf Golosov of the Russian Navy

Numerous books and articles have been written about the leadership evident in the U.S. Submarine Force. The author of one of those writings is Dr. Monroe-Jones, a retired submarine officer. He captured one of the most important defining characteristics of submariners well in an article he wrote for the Naval Submarine League in 2011.

*"Job competence has proven to be the one indispensable factor of submarine leadership. Whatever leadership qualities a submarine commissioned officer or petty officer (terminology for enlisted personnel of some seniority) may have possessed, they counted for little if the person lacked a level of competence that by itself was worthy of admiration."*

Therein lies a key point - Job competence is an indispensable factor of leadership. Write that down. Recall it and test yourself whenever you find yourself in a position of leadership. How many times have you either read about or experienced cases in which a person was identified as a reasonable, maybe even a strong, leader but knew little, possibly even nothing, about the field in which the people being led worked? There are certainly cases in which a person can be a reasonable, even an effective, leader in a field of which that person has very limited detailed knowledge. Many CEOs of major companies and the President of the United States are examples that first come to mind. Setting general direction, controlling finances, motivating the workforce, holding people accountable for adhering to various rules and expectations, and seeing to the welfare of the workers - these are important responsibilities of a leader. They are, however, only a few of a leader's responsibilities, and in many environments, those other responsibilities are just as, and sometimes even more so, essential. A submarine beneath the sea is only one example. Possessing even the highest level of the general skills mentioned above but lacking competence in the job, when the job entails operating and maintaining an underwater war machine, can result in a story with a bad ending. How does one set and reinforce expectations for avoiding detection by those who would do us harm or for maintaining hydraulic systems, electrical components, or nuclear reactors without a strong degree of competence with these items? You can't. When the ship is flooding, sinking, and approaching a depth at which it will implode and kill the entire crew, direction from the leader needs to be more substantive than a good round of cheerleading.

Am I saying that many CEOs of large companies as well as the highest officials in the United States government, are not capable of leadership? Of course not. I'll leave judgment of the competence of those people to others. What I'm talking about here is a special kind of leadership Nuclear Mustang Leadership, the kind of leadership this book is about. The kind of leadership that comes from a person who has either started at the bottom and learned all the aspects of good job performance or, by some other means, has developed the ability to appreciate the challenges and competence of the people working for him. For starters, to practice

Nuclear Mustang Leadership, one has to have job competence. Some readers may be familiar with the term "Mustang," but even those readers might be wondering about the term "Nuclear." Please bear with me. I will share more about this added term shortly.

In a future section, we'll talk about leaders who come into their positions of leadership already having a high level of job competence. But what if you aren't one of those people? What if you don't start with a high level of job competence? How do you get it? Well, starting with the assumption that the selection process for whatever role you are in is at least reasonably effective, you then must have the capability to learn the job and become job-competent, at least to the level of being competent to lead those who are experts in the field. Achieving the level of competence needed involves a lot of personal reading and study, but most importantly, it requires spending time with those who are the experts, the workers in the workplace. I will repeat that essential point: Spend time with the workers in the workplace, learning from them. Get out in the field with them. Hide the authority symbols and go get dirty with them. Ask them questions, lots of questions. Ask a lot about the technical aspects of the job, and the safety aspects of it, but don't limit your questions to these areas. Ask questions about them, their personal lives, what motivates them, and what kinds of problems they deal with at work. Ask them questions that accomplish at least three things: first, make you knowledgeable about what motivates them, what support they need, and how to lead them; second, let them know that you care about them; and finally, and equally if not more so importantly, to learn as much as you possibly can about how to do the job each of them is doing. Sound hard? It is, but do it. Let them train you, and don't let a lack of humility stop you from telling them that. You certainly don't have to, and probably never would under any circumstances, achieve the level of expertise they have, but you will learn what they do, what problems they deal with, how they deal with solving the problems they face, technical as well as manager-caused, the former being just as important if not even more so than the latter.

Let me provide one example of how I once applied the trait of job competency. During one of my tours in the submarine service as a

commissioned officer, I was selected to lead a
highly classified job to correct a submarine
problem that, had it not been corrected, could
have eventually led to the disabling of the entire
nuclear submarine fleet. I would lead a group of
highly competent technicians doing a series of
highly technical tasks. What made this work even
more challenging was that the work was to be
conducted in a highly radioactive environment
with a toxic atmosphere, both of which required
the wearing of not only special and bulky clothes
but also highly cumbersome breathing
apparatuses, as well as severely limiting the time
at the job site because of the radiation exposure
and notably high thermal temperatures
exacerbated by the additional protective gear
required. The first task of the assignment was to
send a specially selected crew of sailors to a
unique training facility that was designed
specifically for this task. Much like the
atmosphere in which the sailors would work, the
training facility required each sailor to don all of
the protective gear mentioned and, wearing this,
to crawl a considerable distance through a small
pipe-like enclosure so narrow in width that it
required one to struggle to even move along in it.
It was like crawling through one of those culverts

under a road, only many times longer, while wearing bulky winter clothing and scuba gear. The first thing I did as the leader of the group was to suit up in all of the cumbersome gear and go through the exact training that each of my crew would have to experience. This move allowed me to understand the challenges that each of them would experience, like the claustrophobic conditions, the heat, and the poor visibility. Because of the training, I in no way came even close to having the level of expertise my crew members developed, but I was able to relate to their problems, better understand the importance of their requests, better able to anticipate their needs, better able to appreciate their dedication to willingly do this job in the first place. It helped make me more competent to lead them. I cannot more strongly emphasize the importance of this trait of Nuclear Mustang Leadership.

In the same article mentioned above on job competence, LCDR Monroe-Jones also included comments from Gordon England (who served as the U.S. Deputy Secretary of Defense under President George W. Bush), who noted several specifics that he said represent the framework of leadership in submarines. These included:

1. Provide an environment for every person to excel.
2. Treat every person with dignity and respect
3. Be forthright, honest, and direct.
4. Improve effectiveness to gain efficiency.
5. Respect the time of others.
6. Identify the critical problems that need solutions for the organization to succeed.
7. Describe complex issues and problems simply.
8. Never stop learning.
9. Encourage constructive criticism.
10. .Identify those who are the greatest performers and delegate to them full authority and responsibility
11. Make ethical standards more important than legal requirements.
12. Strive for team-based wins.
13. Emphasize capability, not organization.
14. Incorporate measures and metrics everywhere.

I would never argue that these specifics do not represent the framework of submarine leadership, but I would argue that these same specifics could be listed by any leader in any organization, regardless of the work they do.

These are not the items that differentiate those in the submarine force from those in any other organization. Find ten or more submariners, ask the leadership question and you'll get ten or more additional specifics or traits. The bottom line is that on a submarine, leadership is not only important, but it is also vital to the survival of every person on that submarine. It is, therefore, not only strong but also somewhat unique. Some traits make submarine leadership special, and it is those traits that are vital components of what in this book is called Nuclear Mustang Leadership. The first is job competency, as discussed above. Another has to do with the seemingly suicidal sailor described in the opening tale.

When the young sailor in the opening tale of this book heard the report that the compartment next to him was flooding with water, he did what any qualified submariner would do when he (or she these days) became aware of a problem. He was not suicidal. He was living the submarine ethos. The problem was flooding. If it was not stopped, everyone, including him, would die. He didn't run away from the problem and let someone else deal with it. What if no one else did? Or could? He ran toward the problem, and he didn't even have to think about doing it. He just did it, like Pavlov's dog, who had been conditioned to salivate at the sound of a bell. The sailor reacted. He didn't hear a bell and salivate, but he sensed a problem and ran toward it. And afterward, there was no big ceremony to recognize the sailor for his brave deed. His was the kind of thing that submariners do. The sailor was meeting expectations. Saving his own life was just a byproduct. The reason that someone would run toward the problem in a case like this is captured well in the quote by Vice Admiral Rudolf Golosov at the beginning of this section. First, know that Vice Admiral Golosov, who passed away in 2022, had received the award of recognition as a Hero of the Soviet Union, and after he retired from the Russian Navy, became a member of the Russian Academy of Natural Sciences. Admiral Golosov said, *On a submarine, the phrase all for one and one for all is not just a slogan, but reality.*

In this quote, replace the terms "on a submarine" with terms like "in our corporation or our company, our organization, or our business." Wouldn't it be great if all of the employees felt the kind of strong ownership of their company

that a submariner feels for his boat? If the boat
fails and dies, he or she fails and dies with it, so
they'll do anything to help it survive. With this
mindset, it is not surprising that a submarine crew
member would run toward a problem, even if it
meant losing his own life. The problem has to be
fixed or the boat and everyone along with it will
die. Again, apply this concept to your business.
Wouldn't it be great to have employees who,
when they become aware of a problem, run
toward that problem? The sailor jumped into the
flooding compartment because he knew if the
flooding wasn't stopped, the boat would die, and
he would die along with it. This is the stuff of
Nuclear Mustang Leadership, the part of the stuff
that comes from the U.S. Submarine Force. Run
toward problems; perform with the viewpoint that
if the company fails, I will fail, and therefore I
won't let it fail. Now we are beginning to build
the formula for, and this element of the formula
is, an aspect of submarine leadership.

This sailor's problem was flooding, but in a
different environment, the problem could be a
dirty living or workspace, a maintenance task that
needs to be done, a leadership vacuum that
screams to be filled, or any number of other jobs

that need to be done. If you saw one of these problems, would you run toward it and take it on, or would you wait for someone else to come along and fix it? If your goal is to be an effective leader, the answer is obvious. But there is more to being such a leader. A lot more. Some of the additional characteristics of a highly effective leader flow from that of "run toward problems." For example, because of the nature of a submarine, small issues can quickly turn into major issues. If a small water leak goes unattended, for example, might it turn into a larger leak? Might it lead to the loss of the submarine? What if it's a small oil leak? Will that lead to a piece of vital equipment seizing because of a lack of lubrication? What if that piece of equipment that seized is the pump that sends fluid to move the planes that control the depth of the submarine? Again, little issues can turn into big problems. The key takeaway here is that this is true in just about any circumstance, not only on submerged submarines. Submariners are taught to question everything going on around them. If it looks unusual in any way, whether it's a leak, an unusual indication, or any other condition different from what one would naturally expect, question it. Not only is this a practice in the submarine force, but it is also an expectation. It is a responsibility of a submariner. As a Commander of the Submarine Force Atlantic once said, "There is not an instant during a tour as a submariner that one can escape the grasp of responsibility." Start thinking like this, and you will be on your way to practicing Nuclear Mustang Leadership.

# A DIFFERENT VIEW OF LEADERSHIP

Emil Mazey, a Detroit Labor Leader and long-time member of the Socialist Party, at a labor meeting in 1946, was challenged as to whether or not he was a Communist, as he purported to be. In his response, he gave us what was to become the long-lived adage of inductive reasoning," If it looks like a duck, and acts like a duck, and sounds like a duck, it probably is a duck." Unfortunately, such inductive reasoning, although it may well apply to ducks, does not apply to leaders. One can look like a leader, act like a leader, and sound like a leader, all as defined in such leadership texts as I critique herein, but these are only trappings of leadership, and they do not mean that one is truly a leader

If people are not following you and doing so in a manner so that you are getting things done, you are not a leader. This seems like an overly basic point, but it is one to remember: True leadership results in getting things done. Leadership is something much deeper than that which can be identified by shallow observations of your words, appearance, and behavior. It is indeed those words, appearances, and behaviors, not alone but rather combined with a set of fundamentals, that define the core of a true leader. These fundamentals are like those fundamental steps of any worthwhile endeavor, the ones "required" by a law of social physics that we will discuss. If leadership were to be characterized by only a few words, appearances, and behaviors, leaders would be cheaper by the bus load. They would be everywhere. But true leadership is not that simple. It requires hard things, like the development of a host of skills and characteristics, many of which require self-discipline, study, practice, hard work, long hours, and often other sacrifices. It is for this reason that true leaders cannot be found in every office corner. True leadership, like excellence itself, as we will discuss, is also something that can be pursued and strived for but never achieved. It is one of those few things that provide great value merely in the striving to achieve it. The remainder of this book is an attempt to describe those not-easy-to-do elements that are at the core of effective leadership. These are fundamentals, not of leadership, but rather of Nuclear Mustang

Leadership. There is more to be said about these fundamentals.

# THE REASON FOR THIS BOOK

In the simplest terms, I wrote this book, as President Clinton once said during a 2004 television interview, in response to a question of why he did a particular thing," because I could." And I can because I have the experience, real field experience, not only as one who has extensively led but just as importantly, as one who has BEEN extensively led. I can also write it because I have a passion for sharing leadership information that I not only have found useful but that I have not been able to find in other sources. More subtly, though, I also have a passion for correcting the insults I sometimes find in texts on leadership. For reasons I have never fully understood, leadership material is often written by consultants or others who have never themselves led. Not infrequently, such material sometimes comes across as insulting or demeaning. I recall one instance in which an author demonstrated a leadership point on how to help someone, using the almost insulting "challenge" of how to deal with your wife asking you to get her a cup of tea. There are so many real-life examples from the workplace that I have used, and even more that could be used, to make the same points, examples that would be so much more meaningful to workers in the field. I have also sensed in some leadership books a class-based disdain for certain levels of the workforce. One author described the clients of consultants as emotionally immature, basing his conclusion on the fact that they had to hire a consultant. I have consulted for more than forty years. The hundreds of clients with whom I worked may have been many things, but being emotionally immature was not one of them.

Much of leadership literature also puts forth only generalities, innocuous enough and sufficiently lacking in detail to be immune to criticism for accuracy not to mention usefulness. For example, the advice offered by one author on how to deal with an employee who has a negative attitude and won't work is "move into the coaching mode." That's all the author had to offer! Other advice, also not unusual, reflects a naivete born of the absence of in-the-trenches experience. One wise advisor counseled with a warning to "not work too hard and leave time for yourself." Of that author, I ask, what does "too hard" mean? 10 hours? 12 hours? 8 intense hours? 15 hours, with much of it being time wasted? In the following pages, I will offer a two-element technique for reducing one's work hours, but my offering will be based on personal experience, and it will have considerably more fruit in its orchard of knowledge. There are other shortfalls I have often found in leadership writings, one I call hearsay experience, in which the person providing the guidance has never experienced the event or situation they are preaching about but rather shares things they have been told. Another shortfall that is even more grievous is thinking that influencing the behavior of other people is a digital act, something one does and then moves on. Many leadership writings, particularly those written by members of the leadership-inexperienced community, are based on interviews with managers. These writings thus pass along primarily hearsay, what they heard another person say. Did he or she really say this? Did they say this and mean something else? Was the intent of what they said altered, even if only slightly, by body language or facial expression? And can one effectively advise another based on what the advisor heard from someone else? Most will know the answer to these questions. Several leadership texts that I have reviewed miss the point that I mentioned earlier, leadership is really about getting things done. I say this because oftentimes so little of these tomes is dedicated to working with subordinates. The most recent of these that I read had 417 pages. Only 30 of those pages dealt with interacting with and influencing subordinates. The remainder dealt with the politics of leading. What does the boss think of me? Who do I need to befriend to advance? How can I look good in the eyes of others? And finally, leadership advice sometimes drifts into that bog of discussions about the difference between leaders and managers. Material is presented in a way that implies executives move digitally between managing and leading and that one can't manage while leading or lead while managing. My suggestion is to forget these two categories. Set the course, then focus your attention on influencing people to move on that course to get things done. Watch over them to ensure they are getting it done, and change their behaviors if they aren't. Do this the right way, and people will do their best, not because you told them to, but because they want to. Throughout this book, you will find the terms leadership and management used interchangeably. I do this because the time spent trying to define the difference between the terms is an academic exercise. It is time wasted and energy wasted as well. James MacGregor Banks, in his book on Leadership, notes that leadership is one of the most observed and least understood phenomena on earth. He refers to a study that turned up 130 definitions of the word leadership. Non-productive efforts, such as trying to define the term, are like hysteresis losses in the electrical circuit of leading. They add nothing to getting the job done. Effective leaders do not waste time or energy, and they know what leadership is about. It's about getting the job done through people. I have seen leaders fail because of poor management skills, and managers fail because of poor leadership skills. There are a set of skills and attributes that are needed by any person in charge to be effective. Whether these fall into one category or another is not worth analyzing. Consider this: A gentleman in Hamelin, Germany, many years ago, played a very special flute or pipe and got all of the rats that had over-run Hamelin to follow him out of town and into the river where they drowned. Contrary to the agreement he had made with the town governance, he was not paid for his service. In revenge, he again played his special pipe and, this time, got all of the children in town to follow him, eventually leading them to a place where they would never be found. Since one conventional description of leadership is getting people to follow you, would this make the Pied Piper a good leader? I doubt it. Also, the objective of one in charge is not to just get people to follow her. It is to get things done. I am reminded of a burly senior petty officer I once faced as a Navy recruit in training. Before starting a training session on seamanship and intending to make it clear that not digesting this training was not an option, he forewarned us with the following statement, "You were brought here to be trained, to drink from the pool of knowledge. Now they say you can lead a horse to water, but you can't make him drink. Well, I can lead a horse to water, and if I stick the business end of a P250 (a high-pressure 250-gallon per minute pump) down his throat, I can make him drink all the f------ water I want. Am I clear?" The only response I can recall was a clearly and loudly voiced, "Yes sir." To this response, the burly senior, offended by the salutation reserved only for officers, then countered, "Don't Sir me you A

-- H---s. I have two parents." Would you define this less–than–th–gentlemanly interchange to be indicative of a failure in leadership or success because he accomplished what he needed to; he got done what needed to be done? He set clear expectations for our engagement and the effort expected of us.

# MUSTANGS

*"Walk a mile in my shoes"*

-Joe South

We are talking here about the leadership traits of Mustangs, but this is not a book about horses. If it were, I wouldn't be writing it. I know nothing about horses other than the term for their leavings can be used to describe some of the extensive writing currently out there on leadership. The attitude reflected in this let's-face-reality opinion is itself a characteristic associated with a different kind of Mustang, a human one. Mustang, the more commonly used term, is used to describe a kind of horse, a horse that has a mind of its own, a wild and spirited animal that is not easily brought into conformance with the performance standards of others, a horse that can be hard to tame and sometimes reverts to its spirited way. But the term is also, in some communities, used to describe a certain kind of person with similar characteristics. A Mustang, with a capital M, is a military officer who was once in the enlisted ranks and then became an officer, someone who has walked in the shoes of the enlisted, someone who understands their perspectives, experiences, and motivation; how an officer's orders will be received; someone who can quickly ascertain when an attempt to avoid an officer's direction is little more than an excuse because he has so often used those excuses himself. A Mustang is quick to see the difference between reality and an environment that doesn't really exist but that might have been born in the minds of some by politically correct necessities, to see that some words from superiors might be driven by agendas not readily apparent to the less experienced or the less wise.

Before moving on, let me ask you to think about the concept of a Mustang. The term comes from the military, but should the concept be limited to the military? Before answering that question, think about what I have said about Mustangs. They have walked in the shoes of those who are doing the work. They are sensitive to what these workers are thinking and feeling. They have a good appreciation for what the workers are experiencing. Why should this be limited to a military environment? It shouldn't.

Mustang attitude is one more element in that leadership formula mentioned earlier. I include this element because a military Mustang typically starts with not a clean plate but a plate with a large helping of the positive stuff that makes being an effective leader much easier. Those in the military who have served in an enlisted capacity and then received an officer's commission often immediately, at least within the military community, garner respect, admiration, and loyalty. Those who perform in any capacity, military or not, can exhibit the traits and actions of a military Mustang. In examining leadership, we'll talk about what traits and actions bring about immediate recognition so they can be most effectively emulated.

I have been a Mustang and received an officer's commission after about ten years in the enlisted ranks where I advanced through the first six seniority grades and was then offered the seventh but turned it down to leave the opportunity for someone else. I wasn't being particularly generous. I just felt that not taking up one of the few grade-7 slots when I was on a path to becoming an officer would have been unfair. With this background, I have my own strong opinions of what works and what doesn't work in leading people. But I also have had the opportunity as a Mustang to associate with numerous other Mustangs. We gravitated to each other because of our similar backgrounds. This grouping tendency is not unusual, in the military or in just about any other business. People naturally seek the company of those with whose values they can readily relate. So, what I present here are not only my feelings and experiences but also those of many other Mustangs as well.

I can say with considerable confidence that a good Mustang doesn't mind getting their hands dirty with those who work for them. They have faith in those workers and trust them because they have a good understanding and appreciation for the extensive capabilities of these enlisted members. Mustangs tend to be more relaxed because their experience allows them to more quickly discern between the important and the less important of the numerous restraints in a military environment. Most Mustangs are particularly humble because they understand what it is like to work for those who often are not humble.

Now, if you know a Mustang and you don't
think he or she is doing a very good job, you
might be right. All men (and, of course, women
as well) might have been created equal, but such
is not the case with Mustangs. Some less-than-
good ones have fallen into a variety of Mustang
traps, like abandoning humility and thinking they
know everything or are superior to their
colleagues who attained a military commission
through another means. But history books also
have their share of Mustangs who have done
remarkably well, like William McKinley, who
enlisted in the Union Army, got a commission,
rose to the rank of major, and then became the
25th President of the United States; Tommy
Franks, the general who enlisted in the U.S.
Army, subsequently received an officer's
commission, was eventually promoted to the rank
of general", earned Afghanistan fame following
the 9-11 event, and was referred to by his troops
as "a soldier's general; and the one-time
Secretary of defense, James Mattis, who enlisted
in the Marine Corps and retired as a general and
Commander of the United States Central
Command. But again, the ranks (forgive the
expression) of Mustangs are not limited to the
military. In 1982, a gentleman by the name of

Rudolph Diesel patented the engine that now shares his name, the diesel engine. As the author of Whatever Happened to Rudolph Diesel, Douglas Brunt describes in that exceptionally interesting book that Mr. Diesel benefitted from a period of sustained membership at each rung in society, from the very bottom to the top. When his son started his first career as a lowly factory apprentice, Mr. Diesel described starting at this level as "good fortune" because the young man would have the opportunity to see the perspective of all social classes.

The leadership formula discussed herein encompasses the concept of good Mustangs, and good Mustangs are all about one thing - getting the job done.

The information presented here should in no way be taken as an intention to denigrate military officers who received their commissions through other routes such as ROTC, military academy, or other officer programs. As is well known, becoming a real leader takes time. It takes making mistakes. It takes years to learn your own leadership style. Most good military officers eventually learn with time. Mustangs just have a leg up on these others, and it comes down to one word. That word is why, as I'll expand upon in a bit, one does not have to be in or have been in the military to practice the leadership traits of a Mustang but first, a few more words about Mustangs.

The lowest rank among U.S. Navy commissioned officers is Ensign, and there are uncountable numbers of jokes and stories about the young, bumbling Ensign, the foolish things their newness in the navy leads to, and how the salty old navy chief has to "mother" the Ensign to keep him or her out of trouble. Much of this undeserved reputation of Navy Ensigns may well have had its birth in the 1960s TV comedy McHale's Navy, which often conveyed the antics of the inept Ensign Parker. When I received my officer's commission and assumed the rank of Ensign, I had already been in the enlisted ranks for about a decade, wore the silver dolphins, indicating I was qualified and experienced in submarines, and had not only completed naval nuclear power training but had also completed a several year tour of duty teaching at that challenging school. I was no Ensign Parker. After I received my officer's commission and received orders to my first submarine as an Ensign, I was assigned to lead a division with a senior chief petty officer, this being an E-8, or the eighth highest grade for enlisted personnel, whose specialty was the same as my enlisted specialty had been, that being electrical operation and maintenance. I had seen and dealt with more electrical equipment problems and been through considerably more enlisted training than the Senior Chief. What did I do? I put my own experience aside until there was a need for it and let the Senior Chief mother me. That was his job, and I let him do it. The ability to effectively facilitate others doing their jobs and, in the process, let them do their jobs is one of the attributes of a Mustang who is also a leader. All worked out well in this case with me and the Senior Chief and we eventually developed a strong professional relationship with equal respect for each other and a recognition of the special attributes each of us brought to the table. Such is the action of a real leader, Mustang or otherwise, because this type of leader, first of all, is humble, and second, has the attribute captured in that one word I refer to above. That word is "experience," not just any experience, but experience on the line, in-field experience, and experience at the working level. He or she has enough of that type of experience to have the confidence to let others take charge, even when one believes they can do it better themselves; the confidence that allows one to be humble and not need to broadcast talents or accomplishments to create an atmosphere in which the development of others can thrive.

The term "experience" is often used but not often understood. If you think you have a lot of experience because you've been in the job for a long time, you're wrong. More on this later. Also, again, don't stop reading because you haven't been in the military. You don't have to have been an enlisted person who moved up into the officer ranks of the military to adhere to the practices of a Mustang. You don't even have to have been in the military any more than you have to have been a president of the United States to emulate the characteristics of Lincoln or been burned at the stake to emulate the courage of Joan of Arc. But there are some things you do need, and these are offered in the coming pages.

I grew up the son of a coal-mining father and a factory-working mother. A good portion of my youth was spent being the kind of fellow most fathers would be reluctant to let their daughters (or their sons) near. My adventures mostly involved petty theft, street fighting, and alcohol before I even turned twelve. But my several run-ins with the law were minor ones, and the positive highlights of my younger life thus ended up being two-fold: I avoided incarceration and I did finish high school, albeit as only a mediocre student. At the insistence of my saintly mother, I made one run at a college degree and attended a local campus of a major university because of the reduced cost, of course. I left after one semester, with a grade point average of 0.3 on a 4.0 scale, failing several courses that were at the high school remedial level to begin with. To add insult to my departure, I stiffed the University for the final bill I owed, which was for the grand sum of $115. Neither my family nor I had the money needed for me to be at the university in the first place. When my dearly loved parents left this world, having lived lives that were too short and too hard, they left me with a significant inheritance - but no money. My inheritance was arthritis. I had no connections, business or political, no money, no motivation, and no special talents or skills. That is how my professional life began, and with that meager set of tools, I drifted with the current of life.

As I look back, I am clear-eyed that my accomplishments in life do not lie in the realm of greatness. I am not a billionaire. I am not recognized throughout the world for any particular accomplishment, invention, or other contribution to society. But I am satisfied with what I have done, particularly given what I started with. I went from a poor, juvenile street punk to a reasonably wealthy, retired senior military officer, business executive, and owner of my own management consulting company. Presented herein is the formulation that allowed me to do this.

I share the above information to support the validity of the point that the advice provided here falls into that often-referred-to category, "If I can do it, you can do it." You can be the kind of a leader a Mustang can be. You can be even more than that. Read on.

# NUCLEAR MUSTANGS

*"America's Nuclear Navy is one of the oldest and largest nuclear organizations in the world and has the world's best safety record of any industry of any kind. (Other than combat) It is safer to work on a U.S. nuclear (ship) than it is to sit at a desk trading stocks."*

-Forbes magazine

As I just said above, you can be more than the kind of a leader a Mustang can be. You can be the kind of a leader a Nuclear Mustang can be. What is a Nuclear Mustang? Someone who has been a Mustang in a nuclear program where the standards for safety and performance exceed those in most industries in the world. Think about it. As stated above, the strengths of leadership provided by Mustangs in any branch of the military are widely recognized, and the strongest element of this kind of leadership, as mentioned earlier, comes from their experience as well as their ability to relate to those being led, to understand what it is like to walk in their shoes. But an even stronger leadership chemical solution results when the traits of a Mustang, are combined with the high operating standards, the natural tendency toward teamwork, and a focus on getting the job done that are the culture of submarine crews. This combination of characteristics could, in itself, be called Mustang Leadership. For the kind of leadership espoused here, however, I have added one more element to that chemical-like solution of leadership, and it is a powerful one - the nuclear element.

As the above quote from Forbes magazine indicates, the U.S. Naval Nuclear Program's performance is exceptional. Recall also that the Institute of Nuclear Power Operations I mentioned earlier, following the Three Mile Island nuclear accident, transferred the most important performance aspects of the Naval Nuclear Program to the civilian nuclear power industry, where performance is now similar. The people who control and maintain nuclear power reactors now have the highest standards of any industry in the world, whether the reactors are used to propel nuclear-powered submarines or are one of those many huge commercial nuclear power plants operating throughout the civilian world. Thus, the nuclear element is a key part of

the kind of leadership encouraged here. Adding
the term Nuclear to Mustang Leadership brings in
the high standards, the professionalism, the focus
on safety and performance, and, most
importantly, the emphasis on striving for
excellence, attributes that exist in nuclear plants,
both military and civilian, have stood the test of
time, and continue to influence worldwide
improvements in the nuclear power industry.

# BRINGING THINGS TOGETHER

*"The whole is greater than the sum of the parts."*

-Unk

Consider the words of the above quote: The whole is greater than the sum of the parts. As indicated, these wise words are attributed by some to Aristotle, but I challenge anyone to find these words in this form having been stated by Aristotle. He is only sometimes identified as the source because he once said something similar. That said, the real source is not important. The point conveyed in the words is exceptionally important and very applicable to the discussion here. Take the strong leadership characteristics associated with submarine life; add the special leadership characteristics common to Mustangs or to any persons who have walked in the shoes of the workers; then add the nuclear element and, in particular, the concept of continually striving for excellence, and one ends up with a form of leadership that is a high-performance, results-oriented, leadership that can accomplish anything, the whole being more than the sum of those three parts. It is this form of leadership that is Nuclear Mustang Leadership, and the remainder of this book will present the remaining parts that make up the whole.

# THE BASICS

*"Here is the prime condition of success: Concentrate your energy, thought, and capital exclusively upon the business in which you are engaged."*

-Andrew Carnegie

Throughout my three careers and more than sixty years of professional life, I have seen many shortfalls in attempts to successfully apply leadership. Two of these shortfalls stand out as recurring. One is a misunderstanding of what leadership is and why it is. Another is the failure to remember that leadership, regardless of the type, requires the performance of certain time-proven, fundamental, and not-always-easy steps. These I call the basics.

The term "leadership" is so often over-used, misused, and maybe not even understood, both in private and government organizations, that it has lost its true meaning. It's a term that is now often either understood to be something it was never intended to mean or is essentially without meaning. To some, leadership has become a term that is frequently used without a thought being given to what it really means. It's often used as a generic fix to problems that no one has a concrete idea about how to fix. When critiquing the failure to resolve an intractable or complex problem, and especially when even the one doing the critiquing doesn't know what to do or how to go about solving it, an often used out is to blame "a lack of leadership." We see this frequently in the media. The statement is a staple in the life of politicians, and it is heard just as often in the business community. The statement implies that if we get better "leadership," the problem will be solved. In the broadest sense, this might be true, but on a practical level, what exactly do we want the newly added leadership to do? Problems aren't solved by "leadership." They're solved by actions. What actions do we want the new leadership to take? If we knew that, we could just direct that those actions be taken and not even bother mentioning the L word. The word "leader" comes from the old English word, laedan, meaning "to guide, bring forth". What if the person you put in place to provide the added leadership did "guide" and did "bring forth" but didn't get the job done, didn't get the problem for which they were hired, fixed? Led people in the wrong direction? Would you applaud their leadership? Not likely. Similarly, if you picked a person for the job who had the commonly accepted traits of a leader, including decisiveness, enthusiasm, integrity, communications skills, loyalty, and charisma, but still didn't get the job done, would you applaud this leadership? The point is this: if we continue to use the nebulous and inconsistently understood term "leadership," we likely will make no more progress in getting important things done than we had before we started using the term.

One of the most frequent reasons for companies invoking the "need for leadership" is to implement organizational change initiatives, yet the record shows that two-thirds of these initiatives fail, and it is commonly known that most companies struggle with execution. A more accurate description of the action referred to as "execution" is "getting things done." So, if you want your company or yourself to be successful, forget about referring to and trying to apply leadership. Add the modifier and refer to and apply Nuclear Mustang Leadership, and that is the topic of this book. Nuclear Mustang Leadership isn't easy; however, one trait of the kind of person who would provide such leadership is a focus on the basics of getting things done. This trait was likely born of necessity in a long steel cylinder hundreds of feet under the ocean. Think about it. If you are in charge of a group of people aboard a U.S. submarine, you likely have somewhere between three or four to several dozen people working for you. But you have no money, and if you did, you would have nowhere to spend it. You can't call a consultant, and if you could, he or she could not get to you and probably would have no idea where you even are. You can't send anyone off to some new and exciting training for the same reason as above. You can think about the problem you're trying to solve, but don't think too long. Problems on a submarine often have an adverse effect on people's lives and need to be fixed now or at least soon, and there is not a lot of time for "thinking about it" or for trial-and-error approaches. Also, searching for the easiest or most convenient way to solve a problem rather than the most effective way is a disservice to the rest of the crew and is not likely to happen. It is these kinds of conditions that accustom one to quickly go to the basics of fixing whatever issue exists since these are usually the only means available anyway. For just about every worthwhile endeavor in life, there are fundamental steps that experience has shown over and over are necessary to reliably and successfully complete that endeavor. Generally, these fundamental steps are not easy to take, and it is for this reason that the would-be leader side-steps or circumvents them and applies other, albeit less effective, techniques. For example, in the endeavor to accumulate wealth, the fundamental steps one needs to take are: generate income, spend less than you earn, invest wisely what you save; and allow time for that investment to build wealth. Warren Buffet is a good example of this. Buffet is one of the richest men in the world, with wealth in the billions. But it wasn't always that way. He began investing at age ten but didn't become rich until he was more than 50 years old thanks to the almost miraculous wonder of compounding dividends and, more importantly, time. These fundamental steps are not easy things to do. Generating income is work; saving is not as much fun as spending, and it takes patience, like Buffet had, and sometimes sacrifice, to allow hard-earned money to sit (in an investment) for any period of time. There are also usually easier, although considerably less effective, ways to pursue most endeavors, and wealth-building is no exception. One can play the lottery and hope for a big win. Good luck with that, but a win is possible. One can look for hot stock tips and gamble through the stock market rather than the casino. Good luck with that also, but again, success with this approach is possible. The internet is replete with get-rich schemes, essentially all of which have an associated downside risk that is considerably high, and most of which are the result of some other person's attempt to acquire wealth without themselves performing those fundamental steps: work, sacrifice, save, invest, and be patient. Apply this thinking to just about any worthy endeavor to accumulate wealth, and you will find that whatever is tried in the final analysis, to reliably and successfully complete that endeavor, one must execute those fundamental steps. It's as if this concept of fundamental steps of an endeavor constitutes one of the laws of what I call "Social Physics." We might call it the First Law of Social Physics.

There is a general tendency in the business world today to continually seek new, more innovative, more efficient, and maybe even easier ways to pursue various business objectives. This sounds great on first hearing; however, there is a pitfall that is too often fallen into while on such innovative journeys. The pitfall is that so much energy is focused on looking for the new technique, the new and innovative approach, the catchy slogan or intended "message" that needed attention is diverted from those fundamental steps that the first Law of Social Physics says are required to reliably and successfully achieve that objective. The new and innovative approach is like a new and shiny object, an ornament, that at first attracts attention but then keeps it, diverting attention from the performance of those social-physics-law-required steps needed for the reliable and successful achievement of that objective. For example, if a company is in the business of selling shoes, then, somehow, it must attract potential clients, convince the clients of the value of a purchase, complete the sale, and then do whatever is needed to get the customer to return to buy their next pair of shoes. These are the social-physics-required rules of the endeavor to sell shoes. If a shoe store manager brings in a newly minted MBA consultant, the manager might soon initiate a search for diversity in his workforce. This could be a good idea, realizing that a sales force with diverse backgrounds might better understand certain customers. He might also institute a personnel safety program to help ensure his sales force remains safe on the job. The consultant might also help the manager recognize that teamwork is important among his workforce and encourage the expansion of employee training to include something like one of those ubiquitous "ropes" courses that all must attend. Finally, in keeping with several leadership guidance sources, the consultant might encourage the manager to exert some effort to make him or herself emotionally vulnerable. Vulnerability goes hand in hand with leadership, these sources say. But at the end of the working day, even though the shoe store now has a highly diverse workforce, working in a safe environment and functioning like a Blue Angels flight team while working for a manager who now shows his emotional vulnerability, if the fundamental steps required of this endeavor, as stated above, are not being performed, and performed well, the store will not be long in existence, because they won't be selling shoes, except by happenstance.

This concept of the fundamental social-physics rule is not a new one. It's an old one said differently by Andrew Carnegie in the opening quote of this section.

Mr. Carnegie could have (but likely didn't) go on to say to the above shoe company manager: You are not engaged in the business of diversity, safety, teamwork, or emotional vulnerability. You are engaged in the business of selling shoes; concentrate on that.

Those who would choose to spearhead an effort to apply Nuclear Mustang Leadership understand and effectively apply this concept of fundamental and necessary steps the basics. So, if you want to be such a leader, the first thing you should do is think about what are the basics of your business, and then critically review them and take whatever actions are needed to execute those basics to the best of your organization's ability.

# AN EVENT WITH LESSONS LEARNED

The plainly dressed lady walked into the room carrying a mop and pail. She set the pail on the floor, carefully dunked the mop into the pail's contents, squeezed out what was an apparent excess of liquid, and began to swab the floor. *Swish, swish, swish*, chanted the mop as it systematically marched back and forth across the seemingly already clean floor. Perhaps motivated by the floor's already apparent cleanliness, the woman made no effort to move any of the tables or other structures in the room but instead steered her mop around the outer edges, giving a wide berth to any obstruction that the mop approached. A small item, about half the size of a golf ball, irregular in shape and reddish-brown in color, lay on the floor in an area not yet reached by the casual mopper. As the mop marched back and forth across the unobstructed sections of the floor, it eventually approached the item and, in one quick *swish*, passed over it, causing it to seemingly disappear, apparently captured by the army of cottony strings that made up the cleaning battalion of the mop head.

The stringy army, directed by the mop pilot, moved on with the task at hand. After about fifteen minutes, when the objective was thought to have been met, the army of strings received their well-deserved rest period, being placed to soak in a bucket of tepid water in the cleaning closet, awaiting the call to their next challenge of cleanliness, unaware of the futility of the plan. Normally, this seemingly inconsequential event would matter to no one. But this was different. This was neither normal nor inconsequential, and it could lead to death.

The above series of events occurred as I watched the cleaning of a room at a large hospital in the United States. The person operating the mop was a well-intentioned, seemingly intelligent person with a great attitude - friendly, polite, and courteous, interested in doing a good job - the kind of person one would enjoy working with or having work for you if you were in a position of authority. I deduced this in the course of a conversation I had with the cleaner following my above observation. The cleaner, who happened to be a middle-aged lady, although that is of no consequence, told me, in response to my questions, that she had never seen her supervisor at any of her workstations and that she had not been told of the importance of her job nor of the reason for that importance. I saw these as fatal flaws in the effectiveness of the cleaner's supervision, and flaws made all the more bothersome because the small reddish-brown object swept up by the mop was a piece of flesh from a human body, now sitting in a bucket of tepid water, further breaking down and waiting to be spread around the next surface chosen for "cleaning."

One of the most important elements of this tale, and it is a true story, is the valuable information I was able to glean from the unfortunate custodial who played a key role in the story. We'll come back to why I was able to get such insights from the lady, but for now, just know that it was because of one of the principles of Nuclear Mustang Leadership.

Although seeming to be a bit threatened when I first began my line of questioning, the cleaning lady soon looked more at ease. I was dressed in nondescript casual but neat and quite professional clothes, so the trappings of a higher-level manager or person of importance were no barrier to communications. I treated her with every bit of the respect with which I would like to have been treated had I been in her position. This required no acting on my part. I did respect her. I had a deep understanding of the importance of her work, evidently deeper than she did, as it turned out. But more importantly, I could relate to her. I had been a cleaning person. I had washed dishes, served food, and cleaned up after people ate, some of whom were inconsiderate of those who would clean up after them. I took out the garbage, wiped the tables, and mopped the floors afterward. I did all this as a "mess cook" on a submarine, an eighteen-year-old enlisted rookie on a submarine full of seasoned seafarers. A mess cook is essentially a kitchen worker and a helper of the cook. It is one of the lowest positions in a submarine crew. But it is a vital position, and any submarine in the United States Submarine Force cannot operate without the mess cooks. If the mess cooks go away, the ability to serve food to the crew goes away. Without the orderly processing of food for the crew, the crew cannot operate, and thus, the submarine cannot operate. Although not so on the highly complex nuclear-driven submarines of today, where specialization of each person's job is needed, in the older diesel-driven boats, a submarine sailor was required to be knowledgeable and able to perform in every position on the boat. Thus, serving for a time as a mess cook was required for a person to complete qualification for the coveted silver dolphins that identified one as a real submariner. It was these circumstances that led to my serving as a mess cook. However, all that said, it did not change the fact that the mess cook is the lowest element of the submarine crew organization. And I was in that position. And it gave me a real sense of what "low" really is, organizationally speaking. And it bothered me when people reminded me of how low in the organization I was. Little did I know at the time that, as I was executing my domestic responsibilities on that old diesel boat, I was cultivating the seeds of leadership and the ability to understand and relate to those at the lower levels of any organization.

Rather than provide a lot more detail upfront on what Nuclear Mustang Leadership is and why I chose the terms, I decided in writing this book to follow the lead of famous historical sculptors and to reveal the concepts and principles gradually. Much like Michelangelo (although by no means would I deign to compare my skills to his) might have taken a big block of stone and knocked away anything that didn't look like David, to, in the end, reveal that final form, I gathered hundreds of examples of supervisor performance that I had experienced, along with other information related to those examples, put them together and then knocked away and discarded any parts that didn't look like the concepts and principles I am trying to convey. Hopefully, this approach will foster a higher interest in the principles themselves and associated actionable items rather than the more esoteric underpinnings of why I selected these points and the terms Nuclear Mustang Leadership. The reader, at the conclusion, will be the best judge of whether or not I accomplished my intent.

So, the first peek at David's hand (or whatever body part you favor comparing this book to) is the principle:

RELATE TO THOSE WORKING FOR YOU, REGARDLESS OF THEIR POSITION IN THE ORGANIZATION.

Recognize there are no unimportant positions, and communicate their importance and the basis of that importance at every opportunity. If you have worked in that position, all the better, but if you have not, exert the effort to develop perspective and compassion - go into their workstations, eat with them, talk with them, listen to them, and understand their likes and dislikes, their challenges. Don't approach them as a king or queen visiting their serfs but as a colleague.

Back to the story of the wayward mopper. What does the story have to do with supervision? Everything. First, realize that as a true story, this provides a real-life example of ineffective supervision with a performance base. That is, a shortfall in supervision that any of us might be guilty of at one time or another and a shortfall in performance that led to a problem. The room that was being cleaned was the operating room at a large hospital, and it was being prepared for the next patient to undergo surgery. To appreciate the significance of this issue, personalize it. Think about how you would react to the situation if that next patient coming into the operating room were your child or one of your parents. The next paragraph will help you to understand why "personalizing" issues is an effective way of

understanding how to react to them - and thus
another principle of Nuclear Mustang Leadership.

Every year, 100,000 people die unnecessarily
in hospitals throughout the United States for
reasons that are preventable and related to
infections, specifically hospital-acquired
infections. In layman's terms, infections can be
thought of as bugs that are someplace where they
should not be. So, people come to hospitals and
unintentionally get bugs on (or in) them and then
die as a result. This fact brings clarity to the
implications of the poor mopping practices
described above. As I said, the small reddish-
brown object that was picked from the floor by
the mop was a piece of flesh from a human body
from the previous operation. It had been caught
up in the tentacles of the mop and then, without
recognition by the cleaner of what was
happening, was smeared, along with its bugs, all
over the operating room floor. All this happened
and was completed just before the next patient
was wheeled into the room to have some portion
of their body carefully sliced open, thus exposing
it to the bugs that were likely to be now on
essentially every part of the floor and whatever
other parts of the room to which they might have

migrated. Additionally, the "soup" of human flesh, bugs, dirt, and whatever now sat in the cleaning closet, allowing the constituent parts to most effectively combine, like simmering soup, awaiting the next "cleaning" assignment, thus providing the potential for this problem to be repeated multiple times, endangering multiple patients, some of whom might be your loved ones. Remember this story and when faced with an issue, personalize it by mentally replacing the potential "victims" of the issue with someone you love, a family member perhaps. Do this to best understand how to react to it. A good leader effectively reacts to issues, and personalizing these issues helps develop the appropriate reaction.

One more example of how personalizing issues helps to better understand how to react to them: I once had an assignment with several large hospitals to essentially coach supervisors with medical backgrounds on how to raise the standards of performance of the medical staff in implementing practices that would better ensure the safety of hospital patients. One of my charges, a young lady whom I will call Beth here, and I were standing in a small room adjoining one of

the hospital operating rooms. It was called a
Ready Room because it was here where a patient
would wait before being taken into the operating
room for a scheduled operation. It is important at
this time to point out that surgery performed on
the wrong part of a patient is an, unfortunately,
continuing occurrence in hospitals, both in the
U.S. and in other countries as well. Even as of
2023, although strides have been made in
reducing the frequency of these wrong-site
surgery events, they continue to happen. In 2022,
wrong-site surgery accounted for 6% of the more
than 1,400 sentinel events reviewed by the Joint
Commission, and be aware that reporting events
to this commission is voluntary, so the actual
number might be even higher.

To address this issue, several protocols have
been developed to ensure the identification of the
correct body part, including marking the part with
an indelible marker before surgery.
Unfortunately, again, this marking practice has
not been fully embraced by some surgeons,
especially those whose self-confidence has been
built on years of experience. As always, a
protocol (or procedure) "in place" but not

IMPLEMENTED does no good - and wrong-site surgeries continue.

Beth and I waited in the Ready Room, standing beside a patient in a mobile bed about to be wheeled into the Operating Room for eye surgery. The patient had a large black X marked prominently over her left eye. In walked a young nurse who seemed a bit surprised and possibly even frustrated by the patient's marking. She left the room and, in short order, returned with a container of cleaning solution and a wipe. Hurriedly, she wiped and wiped the marking until all evidence of it was gone. Intrigued by what we had just witnessed, I asked the nurse what she had done and for what reason. In a voice almost at the level of a whisper, she told us that the performing surgeon "hates to see these markings" because he considers them an insult to his competence to correctly perform a procedure for which he has been highly trained. Beth, as did I, quickly recognized the inappropriateness of this sequence of events and left the scene to discuss it in greater detail. Beth's indignance at what had happened quickly softened during our discussion as she moved into a mode of empathy. The young nurse who wiped away the X was a dedicated caregiver based on Beth's knowledge of her. Beth recognized that in her one-time role as a caregiver, she had also done things that she recognized she should not have done, not out of malice, but for a multitude of reasons: she was tired, overworked, stressed, pressed for time, and other reasons. I had seen this empathy in Beth on several occasions and she would not deny it. She was an otherwise highly competent, highly motivated, hard-working contributor to the hospital staff. I found a solution in her mother.

I knew Beth had a particularly strong relationship with her mother, that she cared for her very much, and that her mother, because of her age, was having some medical problems. I asked Beth if she would be OK with what we saw if it was her mother who was about to be wheeled into the operating room without that extra measure of protection that the surgery would be on the correct eye. I could almost see flames building in her eyes as she vehemently said, "No." I teased the flames by asking what was more important, the welfare of her mother as a patient or the pride of a surgeon who seemed to think mistakes could only happen to others and chose to use fewer precautions for no other reason than to protect his pride. By this time, Beth was almost at a boiling point. In a discussion some weeks later, Beth told me she now consistently visualizes her mother as a patient involved with any inappropriate practice she observes, and she finds her tolerance for avoidable errors considerably lower than it had previously been. Personalizing issues can be effective and is one more element of Nuclear Mustang Leadership.

Now, back to the lady with the mop and the human flesh. As I had walked with the cleaning lady while she left the now bug-laden room I made some small talk to put her at ease and then described to her that I was on a special assignment looking for ways to improve patient safety. When it was apparent that she was comfortable with the discussion and that I had no intention of jeopardizing her job in any way, she became usefully talkative. I asked if she was aware of any expectations regarding cleaning the hospital facilities, and in particular, the operating rooms. I asked if she was aware that this particular facility had a record showing a relatively high number of hospital-acquired infections, higher than many other similar facilities, and that her job could directly impact the number of those infections. I asked if her supervisor had ever come to her as she worked, watched her, and coached her, giving her either positive or negative feedback on how she was performing her work. I asked if she had ever seen her supervisor at any of her work sites. The answers to all of these questions were common, a disappointing but insightful "no."

A conclusion that could be drawn from the above interaction, which went on for more time and discussion than is described here, was that this cleaning lady was a reasonably dedicated employee. However, she had little to no understanding of the importance of the role she played in minimizing hospital-acquired infections, no idea that these infections kill thousands of people every year, and essentially no training for her job and absolutely no coaching on how to best perform her work. So, it should be obvious that the problem was not with her but with her supervisor. Correct? Not exactly.

Because of my position at the time, I was able to ensure the just described occurrence was reported to the hospital CEO, sans names, to protect the one who some would see as the guilty one. The report was made so that at least this specific condition could be corrected.

Before continuing with the storyline of the mopping saga, let me make a side point regarding how I went about seeing what I saw in the operating room clean-up, why I was in there in the first place, and how I went about following up on what I had seen. The answers to these questions lie in *The Observant Eye - Using it to Understand and Improve Performance*, the book I wrote several years ago and that has been found to be useful by many supervisors and managers in a wide range of fields, including power generating facilities, both nuclear and fossil, hospitals, quality oversight organizations, and several other fields where effective oversight becomes even more important because of the level of risk involved in the work. These latter fields include facilities dedicated to helping young people deal with psychological problems and even corporate organizations. I cannot recommend this book more highly, truthfully, because I wrote it, but for two other reasons as well: 1. To my knowledge, it is the only book out there that deals with the topic of observation, and 2. If you are not doing effective observations in your organization, you are neither an effective supervisor nor a leader. I will refer to other aspects of The Observant Eye throughout this book, but to reduce the redundancy for those who have read the book, I will repeat here only the most fundamental points, and then only as needed to present a full picture. I will also not repeat any examples from that book. Now, back to the meat mopper.

Unfortunately, time did not allow for follow-up with the supervisor involved in the Operating Room scenario (or at least the one who should have been involved). But it is a safe bet that this person did not understand the job of a supervisor. If this supervisor worked for you, how would you describe her job to her? Before you answer that question, let me first introduce you to a model I know. This model is also covered in The Observant Eye, but there it is covered from an observation perspective. Here, we will deal with it as a tool for effective supervision.

A question that can be asked about supervisors is, how good are they? A consulting organization I once worked for found it necessary to come up with a reasonably objective way of answering this question: How does one determine whether or not a supervisor is "good"? That is, does he or she perform their job as a supervisor well? In response to this challenge, the question was first expanded to, How good are they AT WHAT? The next logical step was to define WHAT. After much brainstorming on what seemed at first to be a simple question, the consultants and I as part of them came up with the following: It is well recognized that managers and supervisors do many things, but in carrying out their most important function, which is interacting with the workforce, the job of managers and supervisors can be boiled down to the following: They tell people what to do; they then determine if the people did what they were told to do; they make judgments on the quality of what was done; and they then interact with those being supervised to provide feedback, either positive or negative, and then, possibly, further direction. Following this progress along the diagram below continues repetitively, and additional direction is provided, either in that assignment or on the next one. This can be summarized by the following sequence.

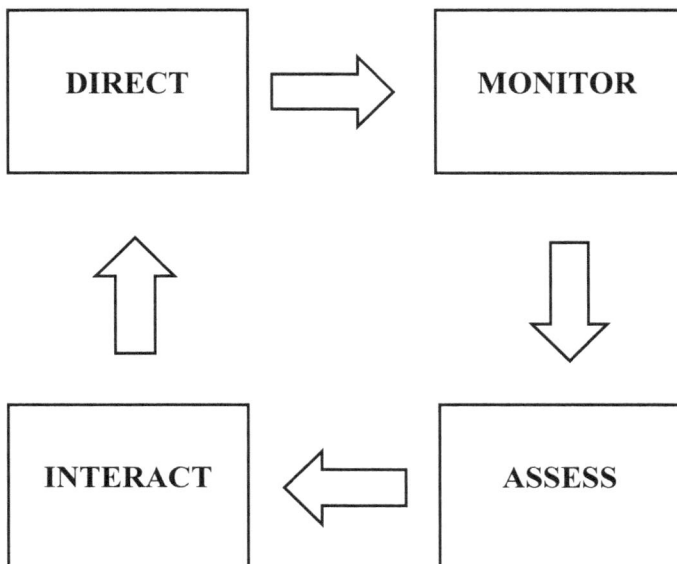

| DIRECT | → | MONITOR |
| INTERACT | ← | ASSESS |

In describing below each of the elements of this simple model I have included a bit of commentary about weaker aspects of each element that I have witnessed during my work with managers and supervisors.

DIRECT -- is the step in which the person in charge tells the worker what to do. In a facilitative work environment, this is still true, except that the input of the workforce is solicited, given fair consideration, and factored into the direction. The description applies as well when workers are "empowered" because management retains the accountability for the final product and consequently de facto gives direction for the work to be done.

The model element of "Direction" is not typically the one that is the primary culprit when a manager or supervisor is weak or not very effective. That said, there is one trap that these people, particularly new supervisors or managers, should be aware not to fall into. It is a trap that I had fallen into during my early days as a manager, so I know it is easy to succumb to. The problem arises when one harbors the often-incorrect conclusion that everyone thinks the same way you do and has the same standards you do. As a result, you assume that a certain direction is not needed when assigning a task because if you were assigned the task, you would perform it to the best of your ability without needing to be told that is how you should do it. For example, if I were told to clean a room, I would like to think that I would approach this relatively menial task the same way I approach any task. I would do the cleaning in such a way that the room would be cleaner than it had ever been. I would mentally compare it with the best-looking room I had ever seen and with that image in mind, would go about my challenge of making this room better than the best. Now, if I assume that everyone thinks the way I do, and I assign someone to clean a room, I would provide minimal direction, probably little more than "clean the room." However, people are different. Each person is a product of their environment and their history of experiences. Some would do just as I would. Some would not. Unless I am confident of the backgrounds and history of those working for me, I should provide sufficient direction such that if the assignee does as I direct, she will have met my expectations. For the relatively unknown assignee, this might include specifying what areas of the room to clean, how to get at some of the areas that are normally not cleaned well; and the importance of and reason for the cleaning. Think back to the meat mopper story and how things might have turned out differently if the custodial had recognized the importance of her work and if someone had emphasized the nooks and crannies likely to hold the "bugs," we were trying to eliminate.

MONITOR is the step in which the manager or supervisor determines if the worker did what she was expected to do. Monitoring can be accomplished by many means, but the primary ones involve going into the work areas and seeing for oneself or reviewing numerical indicators or other written reports.

Monitoring is, without a doubt, the most commonly occurring weakness of would-be supervisors. There is more to be said about this weakness than can be accommodated in this book. It is for that reason I wrote *The Observant Eye,* and the reader is again invited to read that book, preferably before reading this one.

ASSESS (or JUDGE) is the step in which, having seen what the worker has done, the supervisor makes a judgment as to how well the task has been done. Mentally or otherwise, she compares the completed work with other completed work she has seen accomplished or perhaps with work that she has done in earlier years.

I have frequently seen weaknesses related to this element of the model, but not as often as I have of the "Monitoring" element mentioned above. The reason is that the degree of effectiveness in applying this element is directly related to the experience of the individual, and this experience varies so widely among supervisors. A person in charge of anything but unable to most effectively apply judgment to achieve the proper end is likely in this condition because she has not seen the best in whatever she is looking at. Said differently, she does not know what good looks like. This shortcoming is, fortunately, one that can be overcome with a modicum of effort, and the fix is pretty straightforward - to find out what the best looks like. I have had the good fortune to spend the formative years of my professional career in an environment in which excellence in carrying out one's responsibilities was the minimum acceptable - operating a nuclear submarine.

INTERACT is the step in which the supervisor interacts with the worker to give positive feedback if the work is done well; to coach if the best effort was exerted but expectations for final results were not achieved; or to hold accountable and make it clear to the person that his performance did not meet expectations, if that is the response warranted.

Because these are so fundamental to the job of overseeing people and so essential to effective oversight, a principle of Nuclear Mustang Leadership is:

FOLLOW THE MODEL; KNOW HOW YOU MEASURE UP AND WORK TO EXCEL IN EACH STEP - DIRECT, MONITOR, ASSESS, AND INTERACT.

Considering the above basic formula for success makes the job of being a supervisor seem simple. Take these four actions effectively and everything that you want to get done will get done and done well. But I have known many supervisors who have tried this and found such not to be the case. In some cases, when a supervisor tells an employee under her supervision what to do, exerting her best effort to provide direction, the first thing that happens is the employee turns into a disgruntled worker. He feels the direction is not necessary, and thus, the direction in itself is an insult to his capabilities. Sometimes, when supervisors go into the workplace to monitor worker performance, work essentially comes to a standstill as the employees position themselves to avoid trouble that might result from making a mistake while the "boss" is watching. Some supervisors provide feedback that also is not readily accepted, and this, too seems harder than the model indicates it should have been. There must be more to the model -- and there is.

# MANAGER, SUPERVISOR OR SOMETHING ELSE?

Before getting to the more detailed guidance on leadership, a few words about a paradigm that prevents some organizations from getting the most from some of their employees the paradigm of managers and supervisors.

As with many words, the word "supervisor" has its roots in the Latin language. Super or Supra meaning above or over, and vis- meaning see, thus leading to Over-see-or or Overseer or Supervisor. Of course, this is of little help in differentiating among related terms such as first and second-line supervisor, manager, executive, CEO, and Chairman of the Board, since all of those in these positions, at one time or another, oversee something. At least one would hope they do. Dictionaries provide more, albeit little, help. The definition of supervisor in the Cambridge dictionary is someone who supervises. Now that's helpful. An equally unhelpful description is provided in Merriam-Webster: one who supervises, especially an administrative officer in charge of a business, government, or school unit or operation.

The term "manager" also has its roots in Latin. The Latin manus, meaning hand, eventually being picked up by the Italians to form the word maneggiare, which means to handle especially tools. The French word mesnagement and later menagement influenced the development of the meaning of the English word management in the 17th and 18th centuries. The definition of this often-used word is defined in various dictionaries as the act of getting people together to accomplish goals and objectives. Back some time in history, Mary Parker Follett (1868–1933), who wrote on the topic in the early twentieth century, defined management as "the art of getting things done through people."

So, in light of the above, help me out here. What is the difference between a manager and a supervisor in terms of interacting with the employees below them in the organizational structure? I would submit nothing.

Now, some will say that managers have a loftier mission than supervisors, that is, to set and achieve strategic goals, to be responsible for resource allocation and efficient use, and to act in the best interest of the company. Those same people will also see the mission of the supervisor is to interact with the workforce and direct them as necessary to achieve the goals set by the managers. But think about this paradigm for a moment. Who would argue that a manager cannot accomplish anything without interacting with the workforce? And who would argue that a supervisor is not capable of exhibiting ownership of a company, of feeling a sense of responsibility for proper stewardship of resources? The point is that if you operate in the paradigm that differentiates so markedly between managers and supervisors, you are expecting considerably less of both your managers and your supervisors. It is for this reason that in this book, we don't differentiate between the person who fills the position of either a supervisor or a manager. We treat them equally in the discussion. Expect the same from each and you will benefit from a greater focus of the manager on interacting with the workforce and a greater focus of the supervisor on behaving like he owns the company.

# HOW-TOS FOR APPLYING NUCLEAR MUSTANG LEADERSHIP

The remaining sections of this book, to a large degree, are based on the principles discussed to this point but provide additional and more specific advice to those who would implement those principles. This advice is based on what I learned as I progressed from dishwasher to corporate executive, from entry-level recruit to senior military officer in the highly technical field of nuclear submarining, as I monitored and advised workers and executives varying in positions from new naval submariners to commercial nuclear plant operators, to healthcare employees, to those involved in building nuclear weapons, up to and including Chief Executives. It encapsulates specific ways to put into action fundamental principles such as relating to those who work for you, working as hard as you can, learning from your mistakes, and striving for excellence in everything. That advice follows.

# MAKE CORRECTIVE ACTION CORRECTIVE

*"Don't rely on hope. Make sure what you do works."*

-WTS

The following, although told as an analogy to another activity, is a true story. You might find it hard to believe.

There once was a perfectly fine racehorse that was usually ridden by a competent jockey. The owner of the horse continued to be disappointed because of the relatively frequent lack of success of the horse and rider. It seems that, for a variety of reasons, they would occasionally run off the track during races and crash into the crowd that had come to watch the race –not making for a successful racing evolution.

After several such failures, the wise owner, no longer trusting his jockey to properly execute the ride alone, and in a desperate attempt to achieve success (of winning horse races) put a second jockey on the horse to help the first. The owner thought that perhaps the second jockey could help the primary jockey watch the intended track of the steed, and in particular, the edges of the track, and warn the primary jockey if the horse was getting too close to the edge. During the next series of races, on two different occasions, failure struck again as the horse, with its two jockeys firmly in the saddle, went galloping off the track and into the wide-eyed crowd.

The wise owner, who had considerable experience as an owner, was frustrated but, being a problem solver, moved on with another fix to the problem. He would take a highly experienced rider who had risen in the ranks of jockeys to be a manager of sorts and assign him to be a third jockey on the already heavily burdened but determined horse. The owner's thinking was that the primary jockey would ride and direct the horse. The second jockey, as before, would watch where the horse was running and warn the primary jockey if the horse was getting near the edge of the track. The third jockey, applying his maturity, experience, management perspective, and dedication to making sure things were done right for the owner, would ride on the horse with the other two jockeys, watch them, and coach them as needed to ensure they each carried out their responsibilities without deviation.

One day, in the middle of a very important race, the over-burdened horse was struggling down the track when he, albeit at a reduced rate of speed due to his load, suddenly veered off the track and, along with his three riders, plunged deeply into the frightened and screaming crowd.

Shortly after the last event, along came a Nuclear Plant Assessor who had just finished reading *The Observant Eye* for more than the second time. Although he knew nothing about horse racing and little about horses, he decided to stop by the track and sate his curiosity as to why this horse could not be successful, even with the attention and resources poured into the animal and his support, including his three jockeys. The assessor first stopped by the Owner's Area and listened to the sad tale of the multiple failures. He then listened to the owner as he explained why he had put three jockeys on his horse and what he expected each jockey to do. The owner was clearly at wit's end.

The assessor then met with the three jockeys and asked each what they had been doing at precisely the moment when the poor horse went off the track and plunged into the crowd. The primary jockey said he was moving around in the saddle at the time, trying to get a better field of view of the track. He was having difficulty, he said, because there was little room in the saddle since there were two other riders on it with him. He went on to say that because of the positioning problem, he could only clearly see the left side of the track, pointing out that the horse had gone off the right side of the track. When asked what he was doing with the reins at the time, he unhesitatingly, and much to the surprise of the assessor, said that he was not holding the reins. The third jockey, he said, who was a much more experienced rider, he also said, was holding the reins to "help."

The second jockey, when asked the same questions about what he had been doing at the moment of the off-track excursion, also unhesitatingly said he had been watching the right side of the track. He couldn't see it very clearly, nor could he see other parts of either the track or the hands of the rider. Because of his limited view, he said, he didn't realize the horse was turning toward the right side of the track. He, like the primary jockey, also alluded to the limited space in the saddle as contributing to his poor choice for the position he was in at the time of the excursion from the dirt track into Peopleville.

The Nuclear Plant Assessor, understanding that the third jockey was to have been in an oversight role and wondering how, as an overseer, the third jockey could properly execute this role while simultaneously performing some activities of the primary jockey, formulated several questions for the third jockey. The questions tested the overseer's knowledge of what each of the other two jockeys and the determined horse were doing at the time of the unfortunate event. The third jockey's responses indicated he had no idea.

Remembering what the owner had stated were his expectations for each of the jockeys, the Nuclear Plant Assessor got the three jockeys together with the intent of comparing their understanding of the owner's expectations for their actions with what their actions had actually been. The questioning quickly revealed that none of the three jockeys had any idea what the owner's expectations were. The owner they said, had never told them, so they assumed they were being left to their own devices, and their actions were based on what each thought was best.

The above story actually happened. Except, instead of jockeys on a racehorse, it was nuclear workers involved in moving a crane. This analogous horse race was the movement by a nuclear facility crane of equipment highly important to nuclear safety. The primary jockey was the crane operator. The second jockey was a person assigned with the hope that he would ensure the path of movement was always clear and that the crane and the item being moved stayed on that track. The third jockey was a manager, assigned to provide "100% oversight" as an interim corrective action to prevent events that had been happening in which the equipment being moved was inadvertently caused to collide with other equipment.

Again, this really happened, and the lesson learned for anyone performing assessment and/or aspiring to establish Nuclear Mustang Leadership is to not forget to assess the effectiveness of corrective actions, including interim actions, an often overlooked but fruitful target for an attuned assessor. Also, be especially aware of any additional jockeys put on racehorses. Our friends in the Psychology world warn us that putting extra jockeys on racehorses falls into the category of diffusion of responsibility. Simply put, when a task is placed before a group of people, there's a strong tendency for each individual to assume someone else will take responsibility for it so no one does. So, if you see multiple jockeys, ask yourself, what are they doing, and what are they expected to do? The extra jockeys may well be warranted, especially in the kind of evolution where mistakes can be disastrous, but often, the extra jockeys are not fully aware of the expectations of them, and all could be assuming someone else is doing what they, by intent, should be doing. Nuclear power is a field fond of putting multiple jockeys on horses to achieve success, and it is not that unusual when sometimes even this measure, although implemented, is not implemented well.

# STRIVE FOR EXCELLENCE IN EVERYTHING

*"Excellence is a habit and never an accident. It is always the result of high intention, sincere effort, and intelligent execution. It represents the wise choice of many alternatives choice, not chance, determines your destiny."*

-Aristotle

It was March 28, 1979. The time was 4 a.m. Things were happening on a small island that would eventually lead to the spread throughout the United States and, to some degree throughout the world, of a particular type of behavior that was birthed on and below hundreds of feet of water. This type of behavior is only one, but a highly important one, maybe the most important, element of that formula for Nuclear Mustang Leadership.

The island was the Three Mile Island and it sits in the Susquehanna River, just south of Harrisburg, Pennsylvania. Three Mile Island (TMI) is also the name of the nuclear power plant that sits on that island and that underwent an experience resulting in a partial meltdown of the reactor core of one of its two nuclear units. Although no one was injured and no appreciable amount of radioactive material or radiation was released to the public environment, the accident was a turning point in the operation of commercial nuclear power plants throughout the United States. There are wide-ranging opinions regarding both the pros and cons of nuclear power and because it is off-subject, it is not my intent here to support either side of that argument. It is, however, my intent to discuss the tie between Nuclear Mustang Leadership and nuclear power at the time of the Three Mile Island accident. Before this accident occurred, the U.S. Navy had been operating nuclear power plants for more than twenty years without a similar significant problem. The Navy's nuclear-powered journey started when the mooring lines were cast off to allow the beginning of the first cruise of the nuclear-powered submarine, USS Nautilus, on January 17, 1955. The impressive record of the nuclear Navy and the kinds of behaviors that led to this record stood in stark contrast to the traits and behaviors within the commercial nuclear power world at the time.

One of the many action items, and one of the most important ones, taken in our country as a result of the TMI accident, an event that came to be referred to simply as "Three Mile Island," was the formation of a company that would guide and be funded by the commercial nuclear power industry.

The idea for the independent nonprofit company came out of a government commission formed to address the accident - the Kemeny Commission. The company was to be dedicated to operational safety in U.S. nuclear power plants and was called the Institute of Nuclear Power Operations (INPO). The Institute identifies generic safety problems and precursors by reviewing and analyzing nuclear power plant operating experiences. INPO communicates this information to its members to help reduce the possibility of similar occurrences at other plants. It also conducts evaluations of nuclear power plant operations to aid in identifying areas in which improvements can be made. Another of its responsibilities is assisting member utilities in developing and maintaining high-quality, effective training programs. The Institute also aids nuclear utilities in developing and maintaining effective radiological protection and chemistry programs and emergency response capabilities.

A core philosophy of the Institute that is reflected in its work and its expectations of others is the continuing pursuit of excellence, a recognition that one can always do better. Of course, a person cannot expect others to pursue excellence without pursuing it herself, so the newly formed and relatively small Institute, which had about four hundred employees in its early days, was highly selective in hiring only those with the highest personal standards and proven track records of strong performance, particularly regarding nuclear safety. Key players at the Institute thus included U.S. Navy captains and admirals, officers who worked directly for Admiral H.G Rickover, who is known as the Father of the Nuclear Navy, nuclear submarine skippers, commanders of submarine squadrons, and highly experienced commercial nuclear executives, the very people who were instrumental in both the U.S. nuclear navy achieving its high levels of performance and its untarnished nuclear safety record and the commercial nuclear industry eventually achieving its unsurpassed improvement record in nuclear safety following the Three Mile Island accident. I was one of those hired by the Institute after I retired from the submarine force, and these prestigious and highly talented colleagues were the people who served unknowingly as my mentors, the use of unknowing mentors being one of the development techniques used by those in a Nuclear Mustang Leadership environment. I remained with the Institute for almost two decades, starting at the lowest levels of the organization and working my way up to become a vice president and officer of the company. This Mustang theme keeps coming up.

One way to describe the general theme of what I did during my employment at the Institute, as well as in the additional seventeen years of consulting I did afterward, is to say that I inspected nuclear power plants and their management, and in the course of those inspections, looked with a critical eye at whether or not people were truly pursuing performance excellence, with an emphasis on safety, and what level of performance they had achieved. I then determined why they were not if they were not, and developed a report of what needed to be done to more effectively improve. To say that I worked with hundreds of executives, managers, and supervisors in this way would be an understatement. To say that I learned a lot during these interactions would not give justice to the full benefits I received from those I tutored, coached, and sometimes even antagonized. Working with these professionals was the greatest opportunity as well as the greatest privilege of my professional career. This point is captured well in the biblical concept frequently encountered on the internet, but with a source unknown: *show me your friends, and I'll show you your future.*

A key slice of the Institute's mission is improving commercial nuclear power safety by providing highly critical oversight of these operations. A much less advertised part of the mission, especially in its earliest years, and I know this because of my work there from its earliest beginnings, was to essentially transfer the high-performance characteristics, specifically standards and practices, of the U.S. Navy nuclear-powered fleet to the commercial nuclear power industry. In more formal terms, the mission of the company is to promote the highest levels of safety and reliability to promote excellence in the operation of commercial nuclear power plants.

All U.S. nuclear utilities became members of the Institute and as such, were required to commit to strive for excellence in the construction and operation of their nuclear facilities. To assist their "members," INPO developed programs to help utilities in their efforts to achieve excellence. The most visible of these is the INPO evaluation program, in which teams of highly experienced, knowledgeable, and motivated personnel who are committed to excellence in nuclear power operations regularly visit each nuclear plant in the country and evaluate the performance of the workers, managers, programs, and equipment and provide a report with findings that are subsequently followed up on, by the Institute, to ensure the adequacy of their closure.

The first president of The Institute was Vice Admiral Eugene P. Wilkinson, not coincidentally the man who had been the first Commanding Officer of the nuclear submarine U.S.S. Nautilus. Admiral Wilkinson preferred the name "Dennis," and I had the remarkably fortunate opportunity to work for him, having been hired by INPO shortly after its founding in December 1979. It was during my time at the Institute that I came to recognize there is a trait that stands out as the single most important factor differentiating the best leaders from all others. It is a trait broadly recognized as being necessary to achieve the most and to be the best. It is a trait that is contagious. People learn simply from exposure to it. It is the trait of having high standards, believing that everything matters and nothing is unimportant. This trait, which is intended herein to be captured by the term "Nuclear" in front of the title Mustang Leadership, is the true embodiment of a pursuit of excellence. It was from Dennis and his successor, Dr. Zack T. Pate, another nuclear submarine commanding officer, that I learned about this special trait. I would like to think I came out of the submarine force with high standards; however, when I started working at the Institute, I found out what high standards really look like in every conceivable area. For example, my experience at the Institute, among other things, brought me to the firm conviction that a thorough, informative, and well-written report for one of our clients was a flawed report if it had even one typographical error. I also came to better understand that standards of excellence apply as well to so many other aspects of our lives and work, like the way we dress, the civility of our language, our interactions with individuals, and even with our spouses. I also saw that the contagious nature of these high standards positively affected so many of my colleagues as well.

So how did a major industry advance from a level of operational performance that would result in the Three Mile Island accident to a performance level that allowed Bill Gates to say, "Nuclear *energy, in terms of an overall safety record, is better than other energy*"? And it was not only Bill Gates who thought this way. Not at all. Ms. Dixie Lee Ray, a noted environmentalist who is a marine biologist, associate professor at the University of Washington, chaired the Atomic Energy Commission, and was the first female governor of the state of Washington, captured it cleverly when she said: "A nuclear power plant is infinitely safer than eating because 300 people choke to death on food every year." There are essentially no deaths from nuclear power operations, yearly or otherwise. Also, an article in Forbes magazine, as recent as March 2021, was titled "Nuclear Power Continues to Break Records of Safety and Generation."

So, the commercial nuclear power industry had not been performing very well, eventually had a reactor core damaging event, and, by, as some would say, emulating the U.S. Navy's nuclear-powered fleet's behaviors, eventually became a high-performing industry with an admirable safety record. What does this have to do with Nuclear Mustang Leadership? In a word, everything.

That particular type of behavior that I mention in the opening paragraph of this chapter, and that may be the most important facet of Nuclear Mustang Leadership, is the behavior that was conveyed to the commercial nuclear power industry by those from the U.S. Navy's nuclear power world. It is the behavior linked to a striving, almost a passion, for excellence in essentially everything, striving for excellence.

If someone advises you to strive for excellence, what does that mean? How does one strive for excellence? Having worked in military and industrial fields where striving for excellence was routinely a way of doing business and an expectation, I concluded early in my nuclear careers that striving for excellence is a concept that can somewhat be described in words; it can be fostered by pictures, but it is best communicated by example. I have seen commercial nuclear power plants move from the lowest to among the highest levels of performance, and it was clear to me that much of what they accomplished was because they saw what others did, sometimes their own leaders, but often they saw what others were doing, how they behaved. It was these thoughts that convinced me of the usefulness of the concept of exampleship, a concept that is essential to any move toward achieving the best one is capable of achieving. In a bit, I will present a full section on the importance and use of exampleship

As a consultant to the nuclear power industry, my third career and one that followed my retirement from the Institute of Nuclear Power Operations, I was once asked to improve the leadership of a group of six key managers of nuclear plants called Shift Managers. These were individuals whose responsibilities included directly overseeing the field work and operations of those tasked with operating a huge, commercial, multi-unit nuclear power plant.

One of my actions in working with these ladies and gentlemen was to request to be taken to

the deepest, darkest bowels of the power plant. I
knew that in these spaces, the conditions were
difficult to tolerate, and because of that, these
areas were likely to go for longer periods without
being visited and would also, for the same reason,
show the kinds of deficiencies and problems that
would hinder a power plant from achieving the
desired performance. My unspoken message to
these managers was that getting oil or dirt or
sweat on a "pretty" outfit was nothing that any
reasonable washing machine or laundry couldn't
deal with. Then, as we crawled around the plant, I
would ask an uncountable number of questions,
some technical, some managerial, and some of
human interest only. As we crawled about, I
pointed out every single item that I saw them
perform or comment on that was not done to the
best degree that it could be done. By the time I
finished with each one, they were quite
embarrassed, and I was confident would look at
things differently going forward. But they were
not as embarrassed by our walk-around as they
were when I later handed them each a written,
detailed commentary on what they had seen and
done, as contrasted by what they should have
seen and done. Words always seem harsher when
written down since spoken words go into the air

and soon leave no trace of their presence. My direction to each of these managers was to next give these write-ups some thought, then meet with the person to whom they report and tell him or her what they were going to do differently so that such a report could never be written about them again. The exercise was phenomenally effective, and not long after the exercise, the power station received compliments from an external body on their shift leadership efforts. It is important to point out that these managers were not generally poor performers. They were highly intelligent, very capable, motivated people with strong knowledge of their fields. They just didn't know what good looked like, what excellence looks like, and equally important, they didn't know that they didn't know.

The above exercise was a way of encouraging striving for excellence, one more ingredient in the formula for Nuclear Mustang Leadership that becomes even more powerful when combined with those mentioned previously, like:

➤ The Mustang approach of getting into the field and being knowledgeable

➤ Relating to the workers because one has walked in their shoes

➤ Behaviors expected in a submarine environment, such as running toward problems and acting with a motivation driven by the recognition that if the group fails, all will fail.

Information and guidance provided to this point are directed at broad areas; however, there are some more specific recommendations I would make to anyone aspiring to use Nuclear Mustang Leadership, and those are provided in the remaining sections of this book.

# DRESS AND ACT THE PART

*"Those providing Nuclear Mustang Leadership are always setting expectations by what they say, what they do, and what they wear. Speak, act, and dress accordingly. "*

-WTS

Whenever you are in the areas of work, remember that you are continuously communicating, sending messages, numerous messages, even if you are not saying anything to anybody. Be aware of this, dress the part, and act the part. Be professional in your dress, speech, bearing, attitude, and, most importantly, support of the company.

The issue of how formal to dress is not the one I wish to address here. There are numerous articles on the internet discussing how to make such a choice, as well as what the history of "dress for work" is and how the concept of "business casual" came about as well as what it means. If one is interested in whether workers should dress casually or more formally, Sylvie di Giusto, founder of Executive Image Consulting and author of "The Image of Leadership," once provided what I feel is an excellent summary: "The more you deal with a client's money, the more traditional and conservative you should be dressed." Financial businesses would thus have one dress code, and the arts and entertainment business another. I will tell you that none of the above is relevant to the point I want to make here. My point is that dress sends a message. Think about what message you want to send. Shortly thereafter a large company I worked for went to "business casual" my wife and I were on our way to one of our business functions when a very (very) casually dressed gentleman passed in front of our car. It was, in fact, a good friend of mine whose face was hard to see from our vantage point. My wife asked who it was. I told her. She was quite surprised and said, "Oh my goodness! I thought it was the air conditioner repairman". My wife's background is in science and engineering. An accomplished physicist with an advanced engineering degree, she is well accustomed to handling large amounts of money and interacting with a wide range of people at a variety of levels. I value her advice and impressions, and her air conditioning repairman comment drove home a point that is at the root of why I even included this section in this book. The gentleman who passed in front of our car, my friend, was the vice president of one of the major divisions in our company. In this specific example, he was sending a message about who he was. If one were meeting with this gentleman for the first time, on a business matter, is the message, *I am the air conditioner repairman*, the first message you would want someone to receive? As one more example, let me ask what you would think if you were sitting aboard a plane as a passenger and the pilot and co-pilot came on board wearing cut-off jeans, sandals, and tee shirts. What would be your first thought? First thoughts count. In another company, when I was the executive in charge of a large technical facility of several hundred people, I, when on-site, typically wore what one might call business casual, wearing a pair of khakis and an open-neck white dress shirt. But I made a special point of wearing a startlingly white shirt, meticulously pressed, and I shined my shoes each day. My message was and is in telling this story, you will never go wrong by dressing neatly, whatever the type of dress is, and whatever your job.

Sight is not the only important medium in communicating leadership expectations. Sound is another. The messages blaring from a person come from more than one's dress. Speak with poor grammar; use unrestrained profanity; be insulting to any particular individual or group of people, and regardless of your words, you will be saying, "It is OK to do and say what I am doing and saying." If you've had a bad day and want to complain about your boss or the company in general, go ahead and complain. But be aware that there is thus an increased likelihood that your employees will be similarly complaining about you. You told them it was OK to do that when you behaved as you did.

# TAKE CHARGE OF YOUR MIND

*"You won't be able to do it if you think you can't do it."*

-WTS

Training to escape from a sunken submarine is probably one of the last things one would think of as an event that fosters an important aspect of leadership. Think again.

This training, although it can be exciting and even scary, is one of the only experiences I can recall that provides important insight into leadership while at the same time being one that is essentially impossible to duplicate in any other environment. The training consists of a trainee being placed in a small chamber, about twelve inches higher than one's head when standing, and narrow enough that not only is the available room limited to allow only a standing position, but also is narrow enough that one cannot completely stretch out one's arms. The intensity of the experience is magnified by the fact that the trainee will have several other people in the chamber with him or her, thus standing shoulder to shoulder and increasing the likelihood of claustrophobia for those with such leanings. The anxiety factor of the experience increases exponentially as the small, crowded chamber, which is dark except for a single, very small battery-operated light, is quickly flooded with ice-cold water, which rises within the chamber until it is only several inches below the trainee's mouth.

This tight little chamber is mounted near the bottom and on the side of a huge water-filled, silo-shaped tank, with the surface of its water about fifty feet above the escape chamber. A water-tight hatch separates the chamber from the large water silo. After the atmospheric pressure in the chamber is raised to equal that of the 50 feet of water pressure on the other side, the hatch is opened. Trainees, in modern times, wear sophisticated gear that allows them to exit the hatch and "escape" to the water surface above in relative comfort. In certain emergency conditions, when the sophisticated escape gear might not be available, and during my time on the boats when that gear had not yet been developed, one does what is called a "blow and go." This entails wearing no equipment, taking a deep breath, ducking underwater to exit the hatch, and then allowing one's self to rise in the large water column toward the much-desired open spaces above. Like the vacation salesman who convinces you to take the trip that, once embarked upon, makes you wonder if you made the right decision, the trainers tell you during the walk-through of this procedure that you better continuously breathe out during your ascent. Why? Because if you don't, your lungs will explode. The deep breath you took before leaving the chamber is one of highly compressed air created in the chamber to allow the hatch to be opened. Now, as you gently rise in the water column, this air continually expands as the pressure around it decreases, and your lungs expand right along with it. The only way to prevent your lungs from self-detonating is to continuously blow out as your depth decreases. This is not as easy as it might appear on first thought, because of your mind. Your mind is recalling experiences and feeding them into your brain. It remembers those times when you were at home and tried blowing out for as long as you could, for medical tests or contests, or blowing up balloons, or whatever. The mind then continues to do its job and tell your brain, based solely on previous experience, that's all the air there is; now it's time to breathe in; this is the only option you have. Breathing in is certainly the option you want to exercise so badly. But breathing in is not a good option when you are tens of feet underwater, and you know it. So, the successful trainee-cum-escape artist takes charge of her mind, tells it this is not a repeat of the way you've done something similar in the past, but rather, is an entirely new experience, and then continues to push air out of her lungs. And like magic, in an experience new to the mind, the air continues to come out as that in the lungs expands. For safety reasons, two divers accompany each escaping trainee and ensure the air bubbles continually come out on the way up. If at any time the bubbles stop, one diver holds the escaper while the other unceremoniously punches the trainee's stomach to force the air out and, as a reward, gives that would-be submariner one more chance to escape the little chamber. This small chamber fills the role of the escape chamber on a sub, so this exercise is a must-do for any submariner. Only one extra opportunity is allowed before the trainee's submarine career ends before it even begins because of deselection for submarine service.

So, what does this short tale of escape chambers and submariners have to do with leadership? Before answering this question, a few accolades for one of our most valuable leadership tools, the mind. The mind holds our library of operating experience. Among its many other functions, it processes facts in front of us and weighs them against past experiences. It helps us, without our even thinking about it, to calculate the risk of certain actions, what the probability of an event occurring is, and allows us to compare that figure with the potential consequences, based not only on our own experiences but the experiences of others that we have read about or learned about in other ways. So, the mind is an incredible tool, but it can also be a barrier. When faced with a new situation, like being deep underwater, in which you might have no experience to draw on, the mind will draw on whatever it has, and that might not be in your best interest. Taking a deep breath of ice-cold water would be an example.

No one stands out because of the easy decisions they make. They stand out because of the hard decisions they make. Those in charge in a well-implemented Nuclear Mustang Leadership environment are good at making hard decisions, and one of the techniques such leaders can, on occasion, use, sometimes even without thinking about it, is to tell the mind to stand down because this is a new experience. Past experiences that might be of limited use, if any, need to be put aside. The submarine escapee keeps blowing out, and the effective leader decides and moves ahead.

# RECOGNIZE THAT THERE ARE TWO WORLDS

*"Maybe the world has two different kinds of people, and for one kind, the world is this completely logical rice pudding place, and for the other, it's all hit or miss macaroni gratin."*

-Haruki Murikama

Perhaps Mr. Murikama, the acclaimed Japanese author of several international bestsellers, is right, and the world just has two different kinds of people, and maybe those two kinds are those in the workforce who deal with the reality of the macaroni gratin while the managers sit in their offices and dream of the rice pudding.

In many organizations, but not on a submarine, for reasons I'll discuss shortly, there are two worlds: the world of management and the world of those who conduct the business of the business.

The world of management can be a fun place to be for anyone with a vested interest in the company operated by that management. It can be the place, thought by those in it, to be of textbook governance. Management "knows" that people are motivated and dedicated to their work. Those in charge see not problems but opportunities. It is a high-performance environment. At least it is in the eyes of those in it. It is the kind of environment in which management can assume any professional would want to work. There is an obvious, to the point of being assumed, interest in people and in engaging them. There is a recognition of existing problems in the organization, but only in a broad sense. Not a lot of time is spent getting into the details of the problems because that might breed an atmosphere of negativity. A wide range of initiatives and improvement efforts are in place and often under discussion, giving a generally positive feeling about the future. The only problem with this environment is that it is often one in which company success is elusive. This is a fairy tale world, an environment with supportive underpinnings no more likely to exist than the conditions in those tales of the Brothers Grimm. In reality, those underpinnings don't exist. These managers are unknowingly fooling themselves. They're telling each other what they want to hear themselves, and that is, we are doing a really good job.

If you want to know what the real environment is in this place or any other, act like a Mustang. Go into the field, the workplace, where the business of the business gets done. And when I say GO, I mean really go really get involved in the field. There, you will find that it is a different world. It is the real world. The initiatives about which the managers are bragging to each other are not really proceeding as smoothly as those in charge would like to think or are telling their bosses. The managers discuss the concept of excellence and their pursuit of it, but the supervisors in the field are on a different playing field. They are just trying to get the job done, trying to meet schedules, trying to fix broken equipment and dealing with some people with marred if not broken, attitudes. They might have heard of an excellence initiative and badly want to support it, but not right now. For now, they are trying to survive, to just get through their shift.

I have seen more of these two-world scenarios than I can count. I saw them in military organizations, and I saw them time and again when I worked in civilian businesses, one of which I owned. At one facility, in which achieving excellence was, in fact, a top initiative discussed very positively in the management world, I walked out into the field and, within a few minutes, found workers using improper tools, important indication lights that were out of service, and improperly placed tags, tags that were designed to protect people's lives. Practices that were hardly the underpinnings of an environment that reflects a pursuit of excellence.

I have had the opportunity to sit in the management world and witness many meetings, the conduct of which any organization would be proud. The information provided was crisp, clear, and to the point. No time was wasted, and the meetings were thought to be exceptionally useful in defining and advancing the company's objectives. Managers were quick to describe in one way or another their effective meetings as one of the badges of honor of the company. Unfortunately, lower-level meetings, used by supervisors to pass along the directions and values of the company to those responsible for its product, were not so effective. It is only kindness that leads me to use the word "weak" in describing them. They were less than weak. Attendees did other work as the meeting leader spoke. Some caught up on what I assume was badly needed sleep. The crispness and clearness of the manager meetings I had just witnessed were at no time evident, nor was it apparent that most in attendance thought the meeting to be of much importance. I probably needn't tell you that management presence at these meetings was rare at best.

I could go on to describe other aspects of what was happening in the manager world and how it contrasted with that in the working world, but hopefully, I've shared enough to make my point. Intended or not, the message from these companies was that processes are strong and relationships are strong, but results at the only level at which results are achieved, the working level, are weak. These are two different worlds. The only thing that can bring them into synchronism is management presence in the field, the key point of my book, *The Observant Eye.* A good leader doesn't need to be told this. She knows it. She can observe activities in the management world and know that much of what she sees is BS, sometimes posturing to impress the boss, sometimes taking advantage of an opportunity to blow one's horn. She is not misled to believe everything is going as well as described in this environment, and she knows that if she wants to understand what's really happening, she needs to get into the field, and she will get into the field because effective leaders not only know this is where to be, they are comfortable there.

There is another leadership trap associated with this phenomenon of two different worlds, the euphoric feeling that comes to managers who are strangers to the field where the workers are but enamored with programs, processes, and initiatives. A real leader is always alert for an organization that "feels good" while at the same time is not performing well, as indicated by less-than-desired results. The positive feelings of the environment come about as a result of a lot of positive talk, general happiness or, better said, satisfaction of the managers, a generally relaxed environment without pressing timelines or other pressures, and an almost uncountable number of programs and initiatives that gives everyone in the management ranks a warm, pleasurable feeling that things are going well. I recall one organization in this category that bemused themselves at the upper levels with several initiatives, including a heavy emphasis on learning and discussing facilitative management, a good tool in itself for engaging people and aligning them in the same direction to accomplish mutually agreed on goals. At the same time, there was equal emphasis, activity, and discussion around another major company initiative to follow the guidance of those encouraging the use of "crucial conversations," defined by the originators as discussions where stakes are high, opinions vary, and emotions run strong. This may sound like the management world, or as I call it, Happy Management Valley, and it is. But I bring it up here to make a different but equally important point as that of the Tale of Two Worlds. Without even going into the field and observing the environment of the second world, one can sense a problem without leaving the happy valley of management. Listen intently, and you will learn most from what you do not hear discussions of people's performance. Everyone is so caught up in programs and initiatives and the pleasant, albeit deceiving, positive feelings of progress that are not there, or at least not there to the degree they should be, that they have lost focus on what causes good results good behaviors, or, more simply, people performance, or human performance as we will discuss later. Discussions in this somewhat blind organization generally default to talk of the initiatives and the programs and rarely, if ever, move on to people's performance. And how do you change the focus to one on performance or behaviors? You have to first know what current behaviors are! Put down the Crucial Conversation book for a while, set aside your notes on facilitative leadership, get out of your comfortable seat, and get out in the field. The behaviors are there for you to look at if you will open your eyes and ears. Again, I refer the would-be leaders to read *The Observant Eye.* And remember the hierarchy of things: first, BEHAVIORS second, METRICS AND OBSERVATIONS and third, EVENTS. The former two precede the latter, good or bad. If you want to be sure the latter is good, make the first two good. Behaviors are what is happening in your organization, and if you want to be an effective leader, you must know what they are. One can put a value or a grade on behaviors if one has metrics that measure either the behaviors or the results of those behaviors. And finally, if left to their own devices, inappropriate or wrong behaviors, left unchecked, will produce events that will bear witness to those behaviors. An effective leader doesn't wait for these events to occur.

My suggestion for eliminating the phenomenon of having two worlds is simple: get the managers out into the workplace, and don't be afraid to sometimes bring the workers into the world of managers. Invite them to meetings, seek their input on important issues, and give them a venue for sharing their ideas.

I will make one closing point to further strengthen my admonition against a two-world organization, and it stems from my more than two decades in the U.S. Submarine Service. As mentioned earlier, I have had three long-term careers. The first in submarines, the second in civilian commercial nuclear power, and the third in the management consulting business, primarily in technical fields. I did not experience the two-world phenomenon until my last two careers. Why is that? It would be an understatement to say that the space on a submarine is small, so to begin with, there isn't enough space for two worlds. But even more important, the crew of a submarine is a team. Either the entire team succeeds or fails, and failure can take the sub to crush depth and on to the deepest depths of the ocean, where it is left forever. Of course, officer jobs on a submarine are important. But so are the jobs of every person on that submarine. This is no different than in any well-run civilian company. The difference in the submarine community is this understanding of the importance of all jobs, and the communication and teamwork that come when in an environment in which anything but the best teamwork will lead to failure and possibly death. If you want success, think about the workplace, the importance of all jobs, teamwork, and communication.

# PRACTICE EXAMPLESHIP

*"Setting an example is not the main means of influencing others, it is the only means."*

-Albert Einstein

Excellence starts at home. Practice exampleship. I have interacted with far too many supposed leaders who made the sometimes professionally fatal mistake of thinking themselves above the level at which real work gets done. Their thinking seemed to be based on an ill-fated belief that they were too important and working at too high a level to be worried about the details of day-to-day operations. They worried about, and sometimes even acted on, problems that were broad and on a high level. Unfortunately, this often left the smaller problems to fester and eventually grow into bigger problems. Such thinking also limits the ability of the one in charge to interact with those who get the real work done, to appreciate their work, to see their challenges, to ensure their well-being, and, most importantly, by the leader's actions, to set an example for their charges to see.

Much of this book is based on the simple concept that if you want people to do certain things or to act in certain ways, the best way to achieve that is to do those things or act in those ways yourself. Albert Schweitzer said, "The three most important ways to lead people are by example, by example, and by example." Albert Einstein said, as indicated in the opening of this chapter, "Setting an example is not the main means of influencing others. It is the only means." Although Schweitzer was four years older than Einstein, these two gentlemen had a lot in common; most importantly, in my thinking, they were both brilliant individuals. Since they both suggest the importance of setting an example, a wise man would listen.

In the opening story of this book, my sub-mate jumped into the flooding compartment or ran toward the problem to use more business-oriented verbiage. He was leading me, showing me how to act. The others, who didn't panic, didn't worry about themselves but rather focused on saving the submarine and methodically carried out the actions expected of them. They were also leading me, showing me without even thinking about it. The power of exampleship, as I call it, is the strongest, potentially most effective instrument in our kit. Maintain an awareness of this, and especially of the fact that it works even when we don't want it to. Unfortunately, bad examples are also emulated, often to the same degree as good ones.

In short order, I'll discuss the importance of setting consistently high expectations. Accept this for now and realize that a good leader sets the highest of expectations for everything and, most importantly, the highest expectations for his or her own performance. She leads by example. It might be arguable that Ralph Waldo Emerson said the following words, but they ring true in my ear. "What you are speaks so loudly, I can't hear what you are saying." Said in the baser words of a Mustang, if your office would make a pig homesick, don't waste your time telling people to keep their work areas clean and orderly.

My first exposure to the value of example came about five decades ago after I arrived at the Recruit Training Command in the Naval Station Great Lakes, near Chicago. I was about to begin my transformation from a knock-about, going-no-where civilian to a sailor in the U.S. Navy. Shortly after arrival, my universe of existence began to rotate solely around one individual, my Company Commander. The Company Commander is the one who remains with a group of recruits throughout boot camp and instructs them in keeping themselves, their clothing, their equipment, and their living quarters in a manner suited to life aboard ship. He also leads the recruits in military and physical drills to develop their military proficiency and physical stamina. The Navy intends that the Company Commander primarily be an inspiring example of a successful member of the Navy upon whom the recruits can pattern their own lives as sailors and citizens.

Remarkably, I remember nothing of what my Company Commander said. I do, however, remember everything he did. He was the picture-perfect sailor, the definition of "squared away." He was a first-class petty officer, an Engineman. I even remember his name McCoy, Petty Officer McCoy. His uniform almost squeaked it was so clean. It fit perfectly and was neatly creased in all those places it was supposed to be. His hat was blinding white and sat squarely on his head with just enough of a rake to subtly advertise his independence. He talked very little and never joked. He was all business, and I never saw him smile. He understood every navy regulation in the books, believed in them without question, and adhered to them without flaw. He was a walking recruiting poster. He was the sailor that every real sailor wanted to be. The impression was a lasting one. More than fifty years after the fact, I can still recall it. That impression imprinted on my mind the importance, much more than that of words, of setting an example and the impact that one's appearance and behaviors can have on others.

The following is the second part of a story I relayed earlier, one of being tasked, as a highly paid consultant, with improving the leadership capabilities of a group of Shift Managers who were responsible for ensuring the safe and continuous operation of a large nuclear-powered electrical power generation facility. This was decades after my above enlightenment on exampleship. Reflecting on my boot camp experience and the value of exampleship the first thing I did in this assignment was to do what I considered one of the most important things I could do to prepare for this assignment. I got out a steam iron and a can of shoe polish. I then used these leadership tools to prepare my outfit each morning, knowing that I needed to look better than just very good. The process started on the evening before the first day when I ironed a newly purchased and still starched white oxford shirt and crisp set of khakis to achieve knife-edge-like creases. I polished my well-worn steel-toe work boots to what could have passed for a boot camp inspection shine. These managers, by the way, were all men, just by happenstance. Women today also play very important roles at a number of our nuclear power plants, including filling the specific positions of these men I was to work with. I spent a good deal of time the evening before our first meeting, polishing my well-worn work shoes to a high shine and ironing my starched white shirt and clean set of khaki work pants to the point that the creases looked like they might cut a careless and unprotected finger. The following morning, when I met with each of these gentlemen, my first act of leadership coaching was to provide a critique with some hard-hitting feedback but no words. I just stood there in my immaculate outfit and provided each a more than obvious critical-eye look at his pants, shirt, and shoes without comment. My message was clear. If you want those reporting to you to look professional, show them what a professional looks like.

As I provided my visual feedback, I could almost hear what was going on in the minds of a couple of these gentlemen. They assumed I was one of those "pretty boys" from the corporate office, all sparkling clean, who had never seen the inside of an industrial environment before, with its oily surfaces, sweating hot environments, and ultra-high noise levels. At this point, it might be useful to return for a few moments to a previous section of this book that relies heavily on Submariners and to review, in particular, the point made that "Job competence has proven to be the one indispensable factor of submarine leadership." In this assignment, I had job competence. I was quite familiar with the environment of a power plant. I had worked in such spaces for many years. I probably had worked with more of the equipment that we were about to look at than some of these managers. In large part, I was also more than familiar with the surroundings we were about to enter because I had taken many opportunities to get out into the nuclear plants with workers to better understand their challenges and working environment, again as suggested earlier as a Mustang technique.

I then asked each of them, in turn, to accompany me on a tour of the most remote, hottest, dirtiest, oiliest parts of the power plant, the places where many managers wouldn't even think of going, but where I knew from experience that shortfalls in performance, the kinds of deficiencies and problems that would hinder a power plant from achieving that excellent level of performance, would likely be evident. I think it is important to point out that at the time I performed this exercise, I had been a vice president in two different companies an officer in one company, the director of a nationwide effort in another corporate role, and was now the president of my own consulting company. With this background, and especially being fully committed to Nuclear Mustang Leadership, I didn't see myself as at too high a level to get down in the trenches with the workforce. How many executives in your organization would do this?

In this particular effort, only after I had wordlessly expressed my expectations for excellence by setting an example in everything, from dress to plant involvement, and subsequently for performance standards, as I'll describe later, did we even start talking about leadership. By having considerably candid and critical interactions with these gentlemen, I essentially held up a mirror for each to see themselves through my eyes and to be able to self-contrast their standards with mine in all areas.

Later, in the section on Observing, I'll provide more details on the part of this exercise that dealt with performance standards, in which I not only set an example but, equally importantly, showed the power of "observation." Hard to believe, but as you'll read, this part of the exercise was so effective it even resulted in one of the supervisor's wives getting involved.

Exampleship is one of the leading principles because much of what follows relies on the concept of getting others to act the way you want them to act by acting that way yourself. Said another way, lead by example. Said an even different way: excellence starts at home.

As described above, the most useful tool one can have for practicing exampleship is a readily available set of field clothes. These might include a set of neatly pressed khaki slacks, a set of coveralls, a plain cotton shirt, or even an anti-contamination suit designed for working in areas of radioactive contamination. I use the term "field" clothes to include any clothes that those we inspire to lead might wear as they carry out their daily business. It might even be some kind of clothes worn in a field - a real dirt field where agricultural products are born.

To set an example of high performance, one has to be in the arena where that performance takes place, in the field, in the workplace, where the work gets done. I have interacted with far too many leaders who made the sometimes professionally fatal mistake of thinking themselves above the level at which real work gets done. Their thinking seemed to be based on an ill-fated belief that they were too important and working at too high a level to be worried about the details of day-to-day operations. They worried about, and sometimes even acted on, big and broad problems, leaving the small problems to fester and eventually grow into bigger problems. I can walk through a facility, be it a working shop or a corporate office, and in a short time tell if the place is under the "leadership" of one of these ill-advised individuals. I can also identify one of these facilities without even visiting it. Their overall performance will be below that of top-performing organizations that strive for excellence in all they do. What do I mean by striving for excellence? I mean, do everything you do to the highest standard achievable and continually try to get even better. Respond to customers on time; be courteous when answering the phone; treat visitors with the same respect shown to employees and employees with the same respect shown to visitors; and visitors with the same respect you would expect others to show you. Start and end meetings on time don't waste attendee's time; prepare schedules and adhere to them without fault; send no correspondence out with typographical errors because your communications are pictures of you; and this list can go on.

So, what is the role of the leader in establishing an environment in which everyone strives for excellence? A good leader sets the highest of expectations for everything including his or her own performance. Lead by example. While once leading an operating organization of about 800 employees, I had the challenge of getting one of the senior managers reporting to me to get out in the field and see for himself what had happened whenever the facility had an event that led to a shutdown in production. He was reluctant to do this, thinking that, as a high-level manager, he was better suited to remain in his office and have others report to him what had happened. Each time an event occurred, I would go to the scene of the event, ask several questions of those directly involved, and later use this information while quizzing the desk-bound manager about the event. These embarrassing moments for the manager had only to happen a few times before he began to get out to the scene of events, where he soon learned the value that I already knew of getting first-hand information. At the same facility, I was successful in getting the supervisors to up their dress code by appearing in their midst only in a freshly laundered and pressed white shirt that complemented a neatly creased set of work trousers.

At another very large technical facility that deals with nuclear weapons, I had agreed to provide management consulting advice gratis because of the importance of the facility's product to the national security of our country. One late evening, when it must have been obvious that I had spent more than twelve hours in the working spaces, the executive in charge of the facility remarked that I was working particularly hard, and he found that interesting because I was not getting paid. I explained to him that I never considered my compensation whenever I was engaged in work, but rather, that if I agreed to do a job, it was my normal practice to exert whatever effort was necessary to do the job to the utmost best of my capability. He thoughtfully pondered my response, said that was probably why I had been so successful, and didn't even seem to pick up on the fact that I was sending a message to him also. A leader sets an example in everything she does, even when she doesn't realize she is. A real leader sets an example in everything and knows it.

I do not doubt that the most important thing a leader can do is engage the workers and be a role model. During one consulting assignment to a nuclear facility that was trying to raise its standards for nuclear safety, in the course of one day, I counted sixteen examples of missed opportunities to reinforce conservative decision-making. By example, the leaders of this organization were teaching their workforce to tolerate unconservative decision-making.

An effective leader knows that everything she does is setting an example, whether it's apparent or not that people are watching her. And just because the person in charge does something good doesn't necessarily mean that everyone will respond and do that same good thing, but you can bet money that if the person in charge does something not good, like regularly reporting to work late, or not meeting commitment dates, or, as happened in the very unfortunate Uvalde school shooting case, not keeping the security barrier in place by keeping doors locked, these behaviors will be noticed and mentally recorded. These less-than-good performance items will typically get a lot more attention from those you intend to lead than the good things you do. The fix is easy. Assume everything you do is being watched and perform the way you want others to perform.

Set an example in everything you do, and remember John Quincy Adams' word choice on this matter, "If your actions inspire others to dream more, learn more, do more, and become more, you are a leader." He did NOT say If your WORDS inspire. He said if your ACTIONS inspire.

# LEARN FROM MISTAKES

*"Those who fail to learn from history are doomed to repeat it. "*

-Winston Churchill

An effective leader is passionate about learning from mistakes, his own and, just as importantly, the mistakes of others. The submarine force used this principle even in its earliest days, but following the Three Mile Island nuclear accident, the commercial nuclear power industry adopted and then moved this principle to a level worthy of high praise and emulation, and did it in the form of a program called Operating Experience, a program designed to help those interested in doing well to learn from the mistakes of others. Even with this excellent program, the evidence of failure to heed the words of Winston Churchill continues to periodically expose itself. For example, the March 11, 2011, nuclear disaster at the Fukushima Daiichi nuclear station in northeastern Japan, brought about by an earthquake and ensuing tsunami that resulted in the damage of

four of the six reactors at the station, was made
even worse by the company's failure to
implement the lessons learned in emergency
planning during the 1979 nuclear accident at
Three Mile Island nuclear plant in the United
States. Not to make small of the significance of
the Japanese disaster, but more to make the point
that you don't have to have something as
monumental as a nuclear core-damaging event to
understand and take advantage of the value of
applying lessons learned from operating
experience. Neither a submariner, a commercial
nuclear power plant worker, or especially
someone fully committed to Nuclear Mustang
Leadership would ever consider doing any type of
significant evolution without first reviewing and
applying the lessons learned during previous
performances of that evolution. Also, the value of
applying operating experience is just as high in
dealing with even the most basic of activities,
including personal activities, as it is in dealing
with significant technical evolutions. For
example, my wife and I recently made a three-
week, unguided trout fishing trip around central
Idaho and extending into Montana, fishing more
than ten rivers. One of the most important things
we did before departing on the trip was to review

our notes, photos, and recollections from several previous trips we had made around western states and summarize what lessons we had learned from those trips. We then integrated actions to address those lessons learned into our action plans for this trip so that we could benefit from those things done well and not repeat the mistakes we had made earlier. This effort was highly useful, and our latest trip was a complete success and highly enjoyable.

Given that there is considerable value in finding out what kind of problems occurred during a previous run of an evolution in which one is about to engage and then putting measures in place to ensure we are not once again a victim of those same problems, why are we even discussing this in the context of advice on leadership? As with many of the leadership traits discussed herein, there are two leadership aspects of the use of operating experience or lessons learned. These are: first, expectations. An effective leader expects those who work for him to use operating experience, as discussed below, clearly communicates that expectation and holds people accountable for meeting it. The second is exampleship. By one's actions and attitudes, a good leader communicates the importance and value of the use of operating experience.

Finally, as I refer to the concept of using lessons learned or any of the broader submarine and commercial nuclear power concept applications, think about the involved principles in terms of how they can be applied to your professional work, but also to your personal life. If you do, you will be one step closer to establishing a Nuclear Mustang Leadership environment.

# USE OPERATING EXPERIENCE

*"While it is wise to learn from experience, it is wiser to learn from the experience of others. "*

-Rick Warren

Junker Otto Eduard Leopold von Bismarck, a conservative statesman, diplomat, and writer, better known as Otto von Bismarck, is described in the history books as having several character traits, some on the darker side. He is also known for several significant achievements, the best being that of unifying the German states in 1871 to make Germany the most powerful military and diplomatic force in the world. Whether one chooses to admire Otto or not, it is important to recognize that he is the earliest of great leaders to clearly and succinctly, if not somewhat rudely, voice the value of a technique subsequently used to improve performance in the nuclear submarine force, and later in the commercial nuclear power industry. He did this when he said: "Only a fool learns from his own mistakes. The wise man learns from the mistakes of others." This was put in more gentle terms by American pastor and Author, Rick Warren in the introductory quote of this section. Learning from the mistakes of others is the essence of using operating experience to improve performance.

What is operating experience? It can be any kind of experience dealing with the activities of an area. If one is a cook, the related operating experience has to do with such activities as measuring and mixing recipe constituents and applying heat in the process of meal preparation. If one is a school teacher, Operating Experience, or OE, would include experiences interacting with students or other faculty or providing presentations. In a nuclear plant, be it on a submarine or inside of a commercial nuclear power plant, the experience would include starting, stopping, or maintaining the large and often complex pieces of equipment that are used in the generation of electrical power or in the control of the nuclear physics that provides the heat needed for the generation of that power. But the world of operating experience is large, and we need not study it all.

The operative word provided by Mr. von Bismarck is "mistakes." Learn from the mistakes of others. This is the operating experience from which we can most benefit, the mistakes of others. If the cook incorporates home-canned food into his culinary masterpieces and makes the mistake of unknowingly including the bacterium responsible for botulism in his presentation, someone might die. This bacterium, called Clostridium Botulinum, produces a substance known as botulinum, the most toxic substance known to mankind. Any cook can learn from this experience. Using home-canned food? Consider boiling it for ten minutes before consumption, a method known to destroy the bacterium. That would be learning from operating experience.

Just about any field has valuable operating experience. In The War on Error, a book I wrote on the hazards in our healthcare system, I provide numerous examples of hospital operating experience that a patient or a patient's loved one can benefit from during a hospital stay. For example, numerous medical errors due to communication shortfalls resulted in examples such as:

➢ A patient on a ventilator choked to death when prohibited food was provided due to a communication error.

➢ Poor communication led to the death of a dehydrated baby girl who had a hospital-acquired infection and was provided inappropriate narcotics.
➢ A patient who had been moved to an "overflow" room died as a result of a doctor's directions given, assuming the patient would be kept in an emergency room.

Surgery errors have resulted in numerous unnecessary deaths, including:

➢ The wrong rib was removed from a patient with a cancerous lesion on a rib.
➢ A young girl died after receiving a heart and lungs from a donor with an incompatible blood type.
➢ One patient had an unintended cardiac procedure performed on him that was intended for another patient with a similar name.

Improper response to hospital alarms has led to the deaths of a number of patients, including:

➢ A toddler died due to an alarm being turned off

➢  A man died in the intensive care unit when providers failed to respond to breathing and heart rate alarms for an hour.

A patient died when his cardiac monitor alarmed 19 times over two hours, and caregivers silenced the alarms but did not provide needed care.

Any field has a wealth of operating experience, some hard to find, some not. But if you're interested in improving your performance, make the effort. Find out what mistakes have been made by others. Be like Otto von Bismarck and learn from their mistakes.

# FOCUS ON HUMAN PERFORMANCE

*"If, as the person in charge of anything, human performance is not foremost in your mind, you aren't doing your job. "*

-WTS

Human performance, in other words, is nothing more than people doing things. If this seems like a statement of the obvious, it is, but there is a good reason for saying it - again, and again, and again, and again, ad infinitum. And this reason is because it is so obvious that those in charge often don't even think about it, even though they should. Everything that happens in a business happens because people do or don't do something, yet this element of getting people to change their behaviors, and thus to do or not do various things, is usually the element of improvement efforts most frequently missing. It Is not unusual to see get-better efforts include specific behavioral issues such as getting people to wear hard hats, properly complete certain documents, or perform certain equipment procedural steps, but it is unusual to see efforts more broadly oriented toward getting people to perform better in general. I believe this to be the case for two reasons: First, there is a natural human tendency to want to do things that result in clear and readily available evidence that one's efforts have been successful. Need to develop a new procedure? Here it is, in final form. Job done. Need a new program? Here it is program description and all. Again, job done. Do we need to get people to wear hard hats in the plant? Now one can walk around and see everyone wearing the hats. Job done. But what is meant by the statement, I want people to perform better in general? How do I measure that? Importantly, how do I accomplish that? Where do I look for clear and readily available evidence that I accomplished this improvement effort? This is not so easy and so quickly self-satisfying. The second reason for a natural tendency to avoid the area of human performance is that making it better is hard. It requires getting out into the workspace, understanding the problem, and seeing it for one's self.

So, the essential element that is typically missing from improvement efforts is human performance. If you ever see an improvement plan that does not have an element directed at improving human performance, you will be looking at an ineffective plan.

How does one "focus on human performance"? Start by knowing what the level of quality of human performance is in your organization. How do you do that? Start by reading *The Observant Eye. Using it to Understand and Improve Performance.*

It seems appropriate to close out this section by sharing the following. It has been my experience that a common attribute of poor human performance, as evident in a high number of human errors, is a feeling of comfort. When people become comfortable with whatever it is they are performing, the atmosphere is ripe for error. Those in charge in a Nuclear Mustang Leadership environment are never comfortable.

# EXPECT HIGH STANDARDS

*"It's a funny thing about life; if you refuse to accept anything but the best, you very often get it.*
*"*

-W. Somerset Maugham

Hopefully, you won't, but if you ever do yourself begin to wonder about the value of standards, or if you have a young protégé whom you are coaching on the value of standards, recall the adage: *you get what you expect*. If you want excellence, expect it, and as that famous entrepreneur Sam Walton once said, *High expectations is the key to everything.*

In *The Observant Eye*, I make a point about standards, using as an example a directive that each employee is to walk around with a banana on their head. It isn't as silly as it sounds, and a valid point is made by the story. I'll leave the details to anyone interested in checking out that book, but I will make a similar point here because of its importance. The point has to do with standards.

It's not likely that any would-be leader would argue that they need not set high standards, but as with many trite management phrases, like "set high standards," insufficient thought is often given to the words used. What does "high standards" mean? What yardstick should be used to declare that standards are indeed "high"? And what are "standards" anyway? What does the word "standards" mean? Well, the meaning of the word is the first and easiest question to answer. Dictionaries can be of help here, providing descriptors like established for comparison, as a model, as a rule for the measure of quality, and as a basis for judgment. Got it? A standard is a leader's yardstick. So "high" is as high as the leader wants it to be. And, as one expected to lead people, you decide to what your yardstick should be applied. An effective leader would set the bar for the measuring stick as high as possible. She would also apply it, without exception, to everything. I repeat that most important point to everything. For reasons that are likely to be uncovered only through deep psychological analysis, there is a human tendency to apply what I call "situational standards." That is, how high the standards are that we expect to be met is directly related to the importance of the task to which the standard is being applied. If we are doing a very important task, such as operating a highly complex machine for the first time, we tend to be very careful, paying attention to detail and likely following any procedure involved to the letter. High standards are expected in every regard. On the other hand, if the task is perceived to be less important because we have done it many times before, for example, operating our lawnmower, we tend to be considerably less careful. Attention to detail goes out the window and referring to a procedure for such a task would be considered by some to be foolish. But is it really?

Suppose you have a son or daughter who periodically, even if rarely, mows your lawn. And suppose they follow your example and operate the machine without referring to the operating manual. Suppose one day, unaware of the cautions listed in the operating manual, and they accidentally spill some gas on the mower while refueling and then immediately restart the mower and their mowing chore. The Consumer Product Safety Commission has counted at least eleven deaths and 1,200 emergency room visits involving gas can explosions during the pouring

of gasoline. This same commission also reports
that about 60,000 people are treated in hospital
emergency rooms for injuries caused by lawn
tools. My intent here is not to encourage you to
be careful when using lawnmowers. Your
children will be much more effective than I in
describing why they should not be using the
lawnmower. My point is that if you have a
standard or expectation that you, your employees,
or anyone who reports to you refer to an
operating manual when operating equipment,
then that practice should be consistently
implemented. More broadly, if you have an
expectation, be consistent in enforcing that
expectation, regardless of the risk perceived by
others to be involved. If there are certain pieces
of equipment, such as a coffee pot, to which the
standard need not apply, then that should be
stated clearly. The goal is to have consistent
adherence to expectations at the level of
excellence. Leaving the lawnmower and coffee
pot behind, let's bring this point back into the
workplace.

Do not fall into the trap of setting work
standards based on the perceived level of risk or
judging the need for high standards against the

risk if the standards in place are low and could lead to an undesired event. One of the most important things to know about standards is that low standards, not unlike many physical diseases, have a characteristic of creep. They spread. If you accept low standards for anything, and that's another word worth repeating *anything*, then the message you send to those you would lead is that low standards are sometimes OK, and they can select to what they apply the appropriate standard. Low standards are easier to implement, so you can bet on what level the most common standards will be set.

As I point out in *The Observant Eye*, those organizations that have consistent adherence to expectations have been found to experience fewer adverse events, the reason being that people adhere to all of the expectations set by leadership and are not left to decide on their own which expectations to meet and which to ignore. I can still recall my shock when I visited a highly technical facility dealing with nuclear weapons some years ago. The place was dirty! Dirt was on the floor, trash where it wasn't supposed to be. The control of access to one area, in particular, was essentially non-existent. That might not have been bad in itself except for the fact that it was in this area where the most important activities of the facility were performed and where an unnecessary interruption or distraction could have a far-reaching impact, including death. Evidence was replete that hazardous waste was being handled and dealt with in a way that, at best, could be described as cavalier. Standards and expectations existed for every one of these cases, and the standards were reasonably high, as spelled out not only on signage but in policies and procedures as well. So why were the standards not being met? Because the bosses were paying attention to what, in their view, were the most critical activities, and these did not include the aforementioned activities and conditions. There were high standards in this facility, but they were selectively enforced, not on purpose, but rather because the bosses made the bad assumption that if you take care of the important stuff, the rest will take care of itself. Such is rarely the case. If you expect people to do what a sign says, then expect them to do what every sign says. Reinforce a Nuclear Mustang Leadership environment and expect high standards and high levels of performance consistently.

A caution here. I have had more than one experience in which one of my direct reports attempted to use the high standards argument to justify his less-than-stellar performance. The discussion went something like this. Me: Why are the surroundings of yesterday's work area dirty and cluttered? (And it is important to point out that we were not in the janitorial business. Cleaning up our work sites was one of the least consequential things we did. This was a highly technical site where we operated massive nuclear power plants with unforgiving consequences if not operated properly.) Our discussion continued. Employee: I didn't know you expected me to clean the area around the job site after every job. If that was your expectation, you should have told me. Me: Fair comment. Let me be more clear this time. I expect you to do everything you do as well and at as high a level of quality as achievable every time you do it. If you fall short on occasion, we will have a discussion. If we need to have more than one discussion, the second one will likely be of more consequence. If you fall grievously short on even one occasion, there may not be a need for a second discussion. In stating my expectations as I did, I made two things clear. One, we were to strive for excellence. Two, we were to apply this effort to everything we do in or related to our work.

I will say, that as I strive to do with every unfortunate circumstance I experience, I try to learn something from it. In the case of the above gentleman telling me I should have told him in advance what my expectations were, I developed the habit of developing a written set of expectations just about every time I find myself working with a new set of people whose performance I am responsible for. As a consultant in the nuclear power business, I chaired several executive-level nuclear power plant safety oversight committees. For each of these committees, I developed a written set of expectations. These contained many specifics, but among the most important, I included the following:

➢ Safety issues, nuclear, radiological, environmental, and industrial, are to be our top priority.
➢ Facility performance is to be measured against standards of excellence.
➢ We are to provide our clients with a clear understanding of important gaps to excellence in all areas we assess.

➤ The majority of time at any facility is to be spent observing activities and interacting with facility personnel. Paperwork reviews should not be done on-site.

➤ Valuable insights are to be provided whenever possible regarding the results of our reviews.

I mention the above here simply to give examples of what might be covered in expectations, and the reader will note the consistency between these points and the characteristics of a leader in a Nuclear Mustang Leadership environment.

So, as said, setting high expectations or standards is nothing more than expecting people to strive for excellence in everything they do. That is as high a standard as can be set. What do I mean by striving for excellence? I mean, do everything you do to the highest standard possible. It should go without saying that I mean operate highly complex and often even dangerous equipment properly and safely. But it means much more than that. No area of performance is exempt from the expectation. Respond to customers on time; be courteous when answering the phone; treat visitors with the same respect shown to employees and employees with the same respect shown to visitors; start and end meetings on time don't waste attendees' time; prepare schedules and adhere to them without fault; when you finish any work, leave the area in better condition than it was when you started; send no correspondence out with typographical errors, because your communications are pictures of you; be as polite and as helpful to customers as you think possible, and then figure out how to be more polite and helpful, and this list goes on.

As, hopefully, conveyed above, having high standards in every area, important areas, as well as not-so-important areas, is an important element of the environment established by an effective leader. High standards are a way of life, a way of doing business, the right way. How do you get people to have high standards? The answer is, by two means: by setting an example and by expecting high standards of others. The following focuses on the latter.

There is a saying that has its genesis in the early 1960's, "You get what you inspect." Supposedly, this saying has roots in the military, although I was unable to find evidence supporting that conjecture. I suspect that the saying was developed to emphasize looking for yourself or observing activities and conditions to ensure they meet your standards, and I could not agree more with an emphasis on that approach. That said, applying that saying without doing something else beforehand is likely to result in many painful experiences for both the person doing the job and the person in charge. It would be like skipping a step in a procedure, like putting toothpaste on your toothbrush after brushing your teeth. To best appreciate what that missing step is, first recognize that, in general, people want to do the right thing. They really do. So rather than letting them wander off and try to guess what the right or acceptable thing to do is, why not tell them first? Why not let them know what you expect? Then, one could say, as mentioned earlier, you get what you expect.

Striving for excellence is a necessary standard and one that would be set by any leader in a Nuclear Mustang Leadership environment. There are a lot more words to be said about striving for excellence, and it is difficult to talk about standards without talking about the concept of excellence.

At the first company at which I worked as a civilian, a company dedicated to excellence in the operation of commercial nuclear power plants, there was a unique, highly effective, and instructive reminder of that commitment. It was a table, a round table with a glass top. And beneath that top was a graphic display of the muscular forearms of a stone mason working with a mallet and chisel as he carves the word "EXCELLENCE" in a stone slab. The instructional point of this graphic is reflected in the fact that the final E and part of the last C are not yet finished, indicating that excellence is always a work in progress and never finished. Remember, the goal is not the achievement of a state of a thing or place but rather the establishment of the actions and attitudes inherent in the pursuit of excellence. Said another way, as it was by Aristotle, *Excellence is a habit.*

What surprises me about excellence at times is the surprise of other people. I have, on multiple occasions, interacted with those at quite high levels of responsibility in relatively large organizations who were surprised when I shared a story or performed a task that involved striving for excellence. For example, after reading some quite unfavorable press about a large, government-owned nuclear facility, I couldn't help being concerned, as a citizen. As I looked into the unfavorable articles, I concluded that one of the books I had written and mentioned multiple times herein, *The Observant Eye,* could be of use to the organization. So, I sent a copy to the executive management of the facility and soon heard back from them that they had found the book interesting and likely to be of use. To make a longer story shorter, I ended up going to the facility, giving a talk, and demonstrating, by a walk through the workplace, some of the techniques in the book. I did this at no cost to the facility. I felt it was my patriotic duty. During one of the subsequent discussions with executives of the facility at their site, it became apparent to them that I had prepared extensively for my visit. I was familiar with the details of their history, the nature of their problems, and what was likely at the root of their issues. They also noted, based on some of my comments, that I had more than thoroughly prepared my talk and tailored it to their management team, not pulling some previous talk off the shelf but preparing one specifically designed for their purposes. I was asked why I exerted this degree of effort when I was doing this free of charge. My response probably had the greatest positive impact on the organization of anything else I had done with them. I said that this thorough preparation, or doing the best that I could, was just how I do business, and that is, I do not accept every opportunity to perform work, but when I do, I put the same amount of effort (the greatest that I can exert) into providing the best product I am capable of providing, regardless of the compensation or lack of it. And again, that effort is the greatest I am capable of exerting. I firmly believe that if a job, any job, is worth doing, it is worth doing as best as it can possibly be done. They were surprised at my response and appreciative, but more importantly, to my understanding of what this team had to do; their comments indicated this sounded like a new idea they had not thought of before. I share this as another nugget for would-be leaders: *Compensation should play no part in determining the amount of effort that one decides to put into a job one chooses to do - ever.* All I did in the above scenario was to work to my own standard, that being to strive for excellence in all aspects of my work. I don't routinely travel around doing work for no charge, but I would be lying if I didn't admit that the dedication I exhibited in the above case, as well as in the earlier case I mentioned involving my work without charge at a nuclear weapons facility, turned out to be significant future business generators.

Another standards-related characteristic of the kind of leadership discussed in this book is that this striving for excellence must be done by everyone, in every position, in an organization. This characteristic is certainly evident in the crews of U.S. submarines, although frankly, we didn't discuss it using this term. We just lived it.

Dr. Martin Luther King Jr. was one of the wisest men of our time and one deserving of the highest respect. In speaking to a group of high school students several months before his assassination, he encouraged them to have a blueprint for their lives and to have as a basic principle the determination to achieve excellence in whatever field of endeavor was their future. It was during this talk that he made a statement so pertinent to the topic of striving for excellence.

*"If a man is called to be a street sweeper, he should sweep streets even as Michelangelo painted, or Beethoven composed music or Shakespeare wrote poetry. He should sweep*

*streets so well that all the hosts of heaven and earth will pause to say, "Here lived a great street sweeper who did his job well."*

Dr. King might not have known it, but his comments could well have applied to me in my first job on a submarine, a World War II-type diesel submarine. I was not a street sweeper, but, as mentioned earlier, I was the equivalent and did have the lowest level of job on the boat. I was enlisted at just one level above the lowest navy pay grade there is, and I was a mess cook.

A mess cook in the Navy is the one who washes the dishes (we had no automatic dishwasher), cleans the tables and floors after the crew eats, does all of the grunt work involved in feeding a crew, and deals with the garbage. (These days of trying to make everyone feel better about themselves, the term mess cook is no longer used. Now, one is a Food Service Attendant.) Filling this role, as well as every other role on the submarine, was a requirement for qualification to wear the silver dolphins of the enlisted submarine force. This process also reinforced a key theme of the boats everyone is important, and no one is too good to do anything. And again, although we didn't use the word, I

was fully expected to strive for excellence in the
work I did as a mess cook. And why was that?
Because every job on a sub is important. Those
without submarine experience might wonder why
the person washing dishes and dealing with
garbage is important. Bear with me as I digress
and tell a little more about submarines to make a
point.

On a submarine, the garbage can is a
potential killer. It can kill the entire crew and the
boat itself. If you were to study the highly
technical damage control manuals that describe
the design bases and intimate operating
characteristics of a submarine, you would find
that one of the two most catastrophic accidents a
submarine can experience is opening the inner
door of the Trash Disposal Unit, or TDU, at the
same time that the outer door is open. The TDU,
which is a submarine's garbage can, is a tube-like
device, much like a torpedo tube for garbage.
From the TDU, out through the belly of the boat,
are shot canisters of garbage that are weighted so
they go to the bottom of the ocean, unrecoverable
by those who would like to know more about us.
"Shooting the TDU" requires a complex
operation of flood valves, drain valves, and

interlocks. The procedure leaves no room for mistakes. If the operator does not thoughtfully follow the procedure, and a special interlock to prevent such occurrences fails, which is not likely but can happen, both the outer and the inner doors of the TDU open at the same time to create a hole of about 18 inches in diameter. This hole would then connect the warm and comfortable galley (or kitchen) of the submarine to the frigid and highly pressurized water of the sea, thus allowing it to come in with the force of the hundreds of feet of water typically above it. One WW II boat sank off the coast of the U.S. when, in a scenario similar to that just described, both the inner and the outer doors of a torpedo tube were opened at the same time. A torpedo tube opening is close in size to that of a TDU opening. That boat sank to the bottom, with all hands, in 15 seconds. Guess who operates the TDU -- the mess cook. The mess cook is thus provided this opportunity to sink the ship and kill everyone on board every time he takes out the garbage -- about nine times every day.

A person not familiar with submarine operations wouldn't likely know the above. That's not important. What is important is that we recognize that every person's job is important, and as Dr. King advised, everyone in every job should strive for excellence in executing that job. Nothing less than that should be expected of them.

Think about this discussion, and if you aim to establish a Nuclear Mustang Leadership environment, ask yourself a few questions:

Who are the street sweepers (or mess cooks) in your organization?

Are they recognized as important?

Have you told them they're important?

Do they strive for excellence?

Do you expect them to strive for excellence?

Do they know what you expect?

When you've completed this exercise with the street sweepers, work your way up through the entire organization. When you're done, your organization will be a more effective work organization.

Setting high expectations is such an important topic. Please bear with me as I share a few more related thoughts about being consistent in implementing high standards.

An effective leader consistently reinforces the highest standards. On the other hand, as I mentioned earlier, there is a natural human tendency to apply and for those in charge to only reinforce what I call "situational standards" or to reinforce high standards on only rare or special occasions. Why those in charge might choose to perform in this way goes back to a point I made earlier regarding the First Law of Social Physics when I said that for every worthwhile endeavor in life, there are fundamental steps that experience has shown over and over are necessary to reliably and successfully complete that endeavor, and, these fundamental steps are not ones that are easy to take. They are typically hard. The "worthwhile endeavor" we are talking about here is to achieve performance to the highest standards. The most fundamental step to achieve this level of performance is reinforcement of high standards, not occasionally, but consistently every time, in every place. It is much easier for one in charge to not have to correct and coach people every time something is done to a standard lower than it should be, to let it go, maybe even pretend they didn't see the slip. The cost of this is, of course, substandard performance.

In many people's minds, unfortunately, how high the standards are that we expect to be met is directly related to the perceived importance, often time-related, of the task to which the standard is being applied. If we are doing a very important task, such as operating a highly complex machine or maybe starting up a nuclear power reactor, we tend to be very careful, paying attention to detail and likely following applicable procedures to the letter.

On the other hand, if the task is perceived to be less important, like operating an oxygen generator on a submarine or maybe even operating our lawn mower at home, we tend to be considerably less careful. Attention to detail goes out the window and referring to a procedure written specifically for such a task would be considered by some to be foolish. But is it really? I have experienced more oxygen generator explosions than I can recall, and lawnmower accidents cause 85,000 injuries a year, including 70 fatalities. My intent here is not to encourage you to be careful in operating lawn mowers; my intent is to encourage you, when setting an expectation or defining a standard, to be consistent in insisting on its application. My comments here might have sensitized readers to be careful and follow procedures when operating lawnmowers, but what about other lawn-related activities?

Not that I rely on baseball players for professional advice, but it does catch my attention when widely recognized people make statements that clearly communicate a point that I believe is important. In this case, the person is Joe DiMaggio, and he said, "A person always doing his or her best becomes a natural leader, just by example." Notice that this point on exampleship keeps coming up.

Expectations are so important that some organizations put them in a book that is then made readily accessible to employees. Some organizations use a small pamphlet-sized book and require employees to carry it with them during their workday. I have worked in many organizations and seen workers use these books, referring to them frequently. Workers generally want to do the right thing. So, remember what Somerset Maugham once said and what starts this section above, "It's a funny thing about life; if you refuse to accept anything but the best, you very often get it."

# COMMUNICATE CLEARLY, PARTICULARLY FEEDBACK

*"The great enemy of communication, we find, is the illusion of it. "*

-William H. Whyte

Anyone mature enough to be involved in the leadership of an organization realizes the importance of clear communication. It is not my intent here to dwell on that broad and well-covered topic. However, there are related points involving leadership, and it is useful to consider how such communication, particularly as it relates to improving performance, can be viewed in the context of the crew of an operating submarine.

The motivating force behind just about everything done on a nuclear submarine is the goal of successful mission completion within the limited time frame typically allowed for the mission. For example, operating at shallow or even periscope depth for some time without being

detected could well be an important element of a
mission. At one time, as the Ship's Diving
Officer, reporting to me were the Diving Officers
of the Watch (DOOW). These are the people,
usually but not always, officers who, on a rotating
basis, direct the crew members operating the
depth-control planes and control the sub's depth,
as directed by the Officer of the Deck. Among
my responsibilities was to ensure these DOOWs
were trained, qualified, and performed as the
experts they were expected to be. "Broaching"
the sub, exposing it above the surface of the
ocean, or taking it deeper than planned, which
could threaten the mission, were unacceptable.
Time for "gentle" coaching was an unavailable
luxury, so any feedback, particularly that needed
to improve performance, had to be quick, clear,
and effective. Think about feedback activities in
your organization. Are they implemented with a
clear understanding that any corrective action that
needs to be taken based on the feedback needs to
be taken now, and it needs to be, without
exception, effective in fixing whatever the
problem was? Is there a sense of urgency that
ensures the feedback is communicated in a way
that is clearly understood and acted on in a way
that removes any threat to the mission, and there

can be no exceptions to this? If not, read on for a few more words about communication.

One often seen shortfall in the use of words by the leadership of an organization is purposely choosing words that are less painful to the listener to maintain a positive atmosphere. Without thinking about it, maintaining this pleasant atmosphere is given priority over the successful completion of the mission, which, of course, includes completing it within a reasonable timeframe. This can work, but not often, and there is an unintended consequence of its use baseless feelings of comfort regarding the organization's performance and the related unfounded satisfaction that inhibits efforts to improve, to do better. If, for example, you have employees who do not consistently adhere to the company safety policies, the company dress policies, or the policies regarding acceptance and support of employees without regard to gender or other lifestyle choices, that is a "problem." It is not an "opportunity for improvement." It is not a condition accurately captured with the words, "We can do better." The use of such words adds to a kinder, gentler, albeit less effective work environment, and such is often the reason for their use. The downside of their use is two-fold at a minimum. First, any sense of urgency for corrective actions to address the problem or problems is lacking. Why aggressively go after something when it's only an opportunity to do better? Second, and more broadly, the use of such euphemistic words creates an environment that might be kinder and gentler but fosters a general sense of comfort that is not warranted for a substandard situation.

There is another category of non-offensive words, and that is those that have little to no meaning: phrases purposely fabricated to avoid offending or to be unquestionably accurate and non-challengeable to an extreme. For example, I have heard of employees doing something they should not have done or not doing something they should have done, being described as "having lost situational awareness." Really? Does "having lost situational awareness" mean anything to anybody? If the sub you were on is on the surface being shot at or so deep in the water that the communications needed for the mission are lost, would we be discussing situational awareness? Another more subtle example is a statement I read in one company's directive related to emphasizing the importance of human performance. The statement read, "An under-emphasis on the thoughtful use of human performance tools has, in some cases, led to events." Ask yourself, what value do the words "in some cases" add to this statement? What these words do is detract from a concise and clear message.

Would-be leaders would be well advised to remember that those they wish to lead are adults, capable of hearing and processing information that is clear and to the point. Speak clearly, and don't try to make things less than they are, any more than you would have them be more than they are. Do this, and you will communicate the need for action more successfully.

I have been fortunate during my several careers to work in organizations that routinely delivered employee feedback in writing. I treasured the reports that critiqued me. I looked at them as free advice on how to do better and how to advance. I would summarize the key things others thought I should do or not do to be more effective in my job. I kept these written summaries for years, periodically reviewing them to best understand where my self-improvement focus should be. It was in these self-reviews that I learned one of the many important aspects of feedback, and that is, gut feelings about an employee that cannot be based on objective data are often the most useful feedback that a person can get. But that does not mean I always agreed with everything in these summaries. I felt a self-responsibility to improve and advance, and I looked at this feedback as a tool to be used, not blindly, but as part of my overall drive. If a suggestion was made to do something contrary to what I honestly did not think would advance my personal advancement cause, I dealt with it accordingly. For example, in one of my feedback reports, the person responsible for the report said that I was "bloodthirsty and lacked compassion" when critiquing those on whom I provided oversight as they operated large nuclear power reactors. As with all feedback, I took the comment seriously and gave it considerable thought before concluding I really should appreciate the comment, but in a positive sense. I would have preferred the word passionate rather than bloodthirsty, but I took it as a recognition of strength. This comment is an example of the opposite attitude regarding the kind of shortfalls I've described above about being candid when giving feedback. I would pose this question to someone living just down the road from the nuclear plant under discussion: If the people operating this nuclear plant are doing it in a way that could threaten the integrity of the nuclear reactor, would it be preferred that I suggest "opportunities for improvement" to them, or to pat their heads and tell them to not feel bad because I know they are trying their very best? Would any of this kind of communication, particularly since it would not be likely to generate a sense of urgency in getting the performance immediately upgraded, be helpful to the poor person who might have to evacuate their home in the event of a nuclear accident? Enough said. But I will close this part of the discussion with encouragement to any of those desiring to be effective leaders - when critiquing performance, show passion but don't be compassionate.

Another important application of the tool of communication for a leader is, of course, in developing people and, in particular, coaching them. Some companies employ professional coaches, usually highly educated, but unfortunately not often highly experienced, people, often ones with advanced degrees from prestigious institutions of higher learning. I have seen more failures of these types of coaching efforts than I can count from memory. The most frequent cause of these failures has been that no matter what else the coach might know, he or she is not your boss, does not have your boss's temperament, and likely does not have your boss' standards and expectations. Another common weakness of such coaches is they often have extensive consultation experience and even more importantly, recent experience, but, unfortunately, not in the trenches. This point is well articulated by Brene Brown. author of *Dare to Lead; , Brave Work; , Tough Conversations. Whole Hearts.* Ms. Brown says something to the effect, *If you haven't been in the arena getting your ass kicked, I'm not interested in your feedback.* Well said, Ms. Brown.

My advice to companies that hire professional coaches is this. If you want to take concrete action to improve your organization, find the person who hired one of these coaches for one or more of their subordinates and get rid of them or move them to a different lower-level job without responsibilities for developing people. Coaching subordinates is one of, if not the most important activities of a leader. If the quasi-leader is not capable of coaching or thinks she doesn't have time to perform her most important function, there is no hope for her as a leader. You should also look into the failures of your systems that allowed this person to get to such a high level without understanding her role in coaching. An uncountable number of books have been written on coaching, but for an effective leader, it's back to the basic steps required by that First Law of Social Physics: Set clear expectations, ensure they're understood, and give feedback on how well they're met. Give positive feedback to encourage good performance, of course, but keep in mind what Lao Tsu, the Chinese Philosopher and founder of Taoism, once said, "*Great men consider those who point out his faults as his most benevolent teacher.*" Importantly, remember that feedback cannot be any more effective than the communication that relays it.

Just a few more words about communication as it relates to coaching, one of the most important reasons for communications. I'll share my thoughts on this by referring to one of my many dietary shortfalls - salty snacks. My doctor once told me as we were discussing the pros of a healthy diet (unfortunately, there were no cons) it is not the potato chip you eat today that will kill you, but rather the potato chip you ate several years ago. What a great way to communicate that some things that are not good for you have delayed effects. One of those things not good for you, like potato chips, is unclear or, worse yet, misleading feedback. Suzie has been doing a bad job. But you just know that giving her that feedback will crush her. Part of the reason you "know" this is that you just don't want to give her that feedback because it is too difficult for you. This is one of those steps required by that First Law of Social Physics that is hard, too hard, for you. You will feel uncomfortable. You like Suzie. She is a nice person. She has become a friend. She is a great team supporter, and the halo of her supportiveness casts a glow so broadly that it will allow you to rationalize not being honest with her about her current job performance. Also, and unfairly, I might add, without really "knowing" it, you conclude that Suzie is not mature enough to accept candid feedback. So, you lie to her. You come up with some euphemisms, such as those that often start with, "I think it would be good if you …" that, should you end up in a court of law, you can point to as evidence that you did give her feedback. It is not likely you'll end up in a court of law, but there is a very real possibility that one day, when you conclude that Suzie can no longer work for you because she doesn't do a good enough job, and you have to release her, that amongst her teary-eyed response, she might say, "Why didn't you tell me this before? I thought I was doing a good job." The reason such an unfortunate result might come to pass is that way back in the beginning; you gave Suzie potato chip feedback. It certainly felt good to Suzie when it was going down, but in the long term, it was really bad for her. Be alert and refrain from communicating potato chip feedback.

Another thought on effectively communicating on a broader scale has two elements: organization of thought and self-checking. Regarding the organization of thought, an effective leader knows that facts or packets of information are most likely understood and retained if presented in the context of a structure. This is somewhat analogous to the psychological technique of chunking. Suppose one is given ten single digits, in a sequence, to remember, such as 7809814590; some, although not all, will have difficulty achieving a high degree of success in accomplishing this. But, if one breaks these digits down, or "chunks" them, into 780 981 4590, then the task will be easier because now there are only three items to remember. So, it is similar to bits of information provided to someone else if they are provided in the context of a structure, much like holiday ornaments hung on a tree. The provider first provides the structure or the tree and then the individual ornaments. For example, let's say I observed John leading a production-oriented meeting, and I decided I should give him some constructive feedback. Let's say I saw the following four things that John did or said during the meeting that bothered me because they were not done as well as they could have been.

1.  He covered metrics, the values of which indicated a project was considerably behind schedule, and for each deficient metric, it was explained that the reason lay in a faulty piece of equipment that had either broken down or not performed as expected.

2.  One of John's questions went unanswered when Joan, an appointed attendee of the meeting, was unable to attend the meeting, albeit for a good reason, and Sally attended in

her stead but did not have the knowledge level of the questioned subject that Joan had, and was therefore unable to answer the question posed.

3.  Several attendees commented during the meeting, with obvious frustration, that multiple jobs were recently delayed because the administrative procedure controlling the work being discussed is unclear and subject to various interpretations.

4.  The meeting ended with a discussion of some important work that had been completed. During the discussion, a supervisor interrupted and apologized because the work had not been completed, but was only reported as such by him because that was what he had been told earlier.

If by the time John hears the examples of what led you to the conclusion that he has a problem, he understands what the general message is, he will be more willing to listen to and understand the examples provided. There should be no misunderstanding that he needs to perform differently in the future.

To address the final thought on this topic, i.e., self-checking, let's go back to the opening quote of that great Sociologist, William H. Whyte,

>  *"The great enemy of communication, we find, is the illusion of it."*

The act of communicating on a submarine, or, I would argue, in any high-risk occupation, is one of the most important activities in which we engage. On submarines, we valued it so highly that we employed what we called "three-way communication." The process went like this: I give a direction, it might be, for example, make your depth 400 feet. The receiver of the communication repeats what he or she thought they heard, as in, "Understand, make my depth 400 feet". If the words received are, in fact, the words I communicate, I close this communication event by saying, "That is correct." If the words are not what I intended, the sender says, "Wrong." And then repeats what he intended to send. I am not encouraging the use of three-way communication during discussions with others. I only describe the three-way technique to emphasize the importance placed on certain activities when verbal transmissions are important. I am also encouraging anyone involved in providing someone feedback or a critique, which is certainly one of the most important activities in which a leader can engage, to use some method of ensuring that the communication was clearly understood. My method is pretty straightforward. After providing the comments to whomever I am providing them to, I straightforwardly ask, do you understand what I just told you? Tell me what you think I told you. Remember, the bottom line is to have communication, not the illusion of it.

# SEE THROUGH THE BS

*"Bullshit is a greater enemy of the truth than lies are. "*

-Harry Frankfurt

I have been blessed in my professional career to have been associated with a large number of the most highly effective leaders. Often these were people who had filled senior leadership positions in the U.S. Naval Nuclear Submarine Force and then went on to successfully lead commercial nuclear power organizations. One of the common traits of these special leaders that I tried hard to emulate, and that, hopefully, I was successful in doing so, was being able to "see through BS." Harry Frankfurt, the great American philosopher, opened this section with his views on the danger of BS, and It would be incomplete to move on from the subject of communication without saying at least a few words about it. Stating the obvious, managers and leaders often receive briefs on various topics, the most important of which are often problem status reporting, although briefings might also be on the status of more routine matters. In either case, the leader often uses information from these briefs to either take or direct others to take action, and so, again stating the obvious; it is important that the information received in these briefs is honest, candid, accurate, and to the point. Unfortunately, the less experienced person, whether a manager or not, often tends to report status or describe conditions or actions in terms that, although they might make the person seem more competent or successful, do not convey the most accurate description of the situation and thus limit the ability of the more senior person receiving the briefing to apply her experience and expertise to most effectively improve the situation. A good leader does not allow such a condition to develop. He or she can truly understand the briefer, or maybe better stated, "see through" the inaccurate or non-useful information. For example, one guiding principle on this topic, which applies to problem discussions, is that the discussion of actions is the fog that obscures results. Remember this point when one of your reports is telling you about all that has been or is being done, with little to no mention of results achieved.

The ability to most effectively understand what is being presented is the main topic of this section. This ability is a trait common among effective leaders and one that should be emulated by all who aspire to be an effective leader. It is not some magical trait or ability that a certain person has come by naturally, but rather a skill that they have developed, and one that anyone, with sufficient effort and time, can also develop. It is an ability with two main constituents: First, experience, the kind of experience that builds the urge within the listener to say, to themselves, if not out loud, "I've heard this line before," or, "I've been there; done that." The second constituent is the deep knowledge of the person providing the briefing. As an example of a simplified application of these constituents of experience and deep knowledge of the person speaking, think about one's spouse or one's parents. It is possible but difficult, particularly after some years together, to BS these people. This is primarily because they have so much experience in the fields typically being discussed, i.e., the fields of life, and they have such deep knowledge of us because they have seen us do and say so many things under so many different conditions and circumstances.

Returning to the leadership realm, here are a few more words about deep knowledge. The person aspiring to be a good leader has a responsibility to acquire that deep knowledge about those who report to him or her, to get to really know them, both personally and professionally. This acquisition is best done by engagement with the employee, primarily but not limited to spending time in the workplace with them, watching them carry out their responsibilities, and talking to them to fully understand what they are doing, especially performance-wise, and why.

I should also point out that the employee in one of these interactions also has a responsibility to ensure to the best of their ability that their leader acquires that deep knowledge of him/her. This responsibility can be effectively executed by sharing perspectives, demonstrating decision-making, and even challenging comments from the more senior person when appropriate, generally just letting themselves be seen for what they really are, and most importantly, never trying to hide or cover up anything or make things sound even a little bit better than they actually are. An effective leader would clearly communicate his or her expectations regarding this behavior and then regularly encourage it.

I've purposely expended a lot more words on the trait of seeking deep knowledge of the people, more so than on the experience factor, because many senior leaders have experience, but it is the degree to which this second item, deep knowledge acquisition, is carried out, that differentiates the best senior leaders from those that are just good.

I have been fortunate to have had the privilege of working in companies that could be models for what to do and how to do it regarding the above discussion, but I have also observed and experienced the other end of the spectrum as well, particularly in the Navy, outside the submarine community I should say, where those in charge were not always the most experienced, the motivation to acquire deep knowledge of the workforce was often minimal, and the efforts to effectively engage was nil. I have had similar impressions also in several civilian fields. I mention this to prepare the future leader to not expect others to perform in the ways described here. They might, and then again, they might not.

It will be a part of the leader's role to help those future leaders develop in this regard.

# SPEND TIME IN THE GARDEN (OF NUCLEAR MUSTANG LEADERSHIP)

*"Not every difficult and dangerous thing is suitable for training, but only that which is conducive to success in achieving the object of our effort."*

-Epictetus

A nuclear submarine is one of the most complex machines on the planet. Get on one, and you will find that on just about every square foot of surface on which something can be mounted, something is mounted, something of use. You will unlikely be able to stretch out your arms and not touch a valve, a gauge, an indicator, a weapon, a communications device, or an emergency use item, many of which you will use on a not-infrequent basis. You will be surrounded by pumps, motors, equipment to reduce carbon dioxide in the air, others to generate the oxygen needed to breathe, sonar equipment, and more hydraulically operated items than some U.S.

politicians have excuses, all of which must be
maintained operating or ready for operation at
any time. Now, put this complex environment in
a large metal capsule-shaped container and put it
hundreds of feet underwater, essentially cutting
off all communication between those in the
container and those in the outside world, and have
it stay there for several months. While in the
container, get involved in activities, any one of
which would send shivers down the spines of
most normal human beings, including fighting
fires, stopping flooding, dealing with the loss of
ability to breathe, and recovering from
uncontrolled transients to depths from which the
sub might not be able to recover, all while
engaging enemies of our country doing similar
things on other submarines while also preparing
to kill you. What is the second word you think of,
assuming the first is either crazy or exceptionally
motivated? (I prefer the latter; however, a study
of the personality characteristics of successful
navy submarine personnel conducted in 1996
found that 37 percent of those tested met the test
criteria for a personality disorder.) That second
word many would think of is the word "training."
Epictetus, the Greek philosopher, had it right in
the opening quote of this section, essentially

saying one need only train on those things important to accomplishing whatever our objective is. Those are a lot of words easily replaced, at least in a submarine environment, by the one word, "everything." The U.S. submarine force is all about training - training, more training, more training, and then more training. Everything you do, you will be trained to do. Want to use the restroom? You will be trained on how to operate the several valves and the lever needed to flush the commode. Want to throw the garbage out? You will be trained on and then assessed for your ability by formal qualification to use the device that expels trash from the sub. Want to take a shower? You will be trained on how to take a submarine shower, which is a euphemism for a not-very-comfortable washing that achieves its objectives of getting you marginally clean while, most importantly, conserving the limited supply of fresh water. Why do I tell you this? To make the point that training is a part of submarine life. One trains on these things because one's life depends on their proper use. Think about some time when you might have been trying to train some unmotivated young person in your organization on how to do something. Think about how things might have gone differently if you could have said truthfully, "Pay attention to this, or you might die." Training is in the blood of the submariner. But more relevant to the discussion here is the fact that training is the garden in which Nuclear Mustang Leadership grows. It can also be the place where it meets its death.

I am a devotee of training, and my involvement with it has been extensive. I went through the typical training pipeline of the submarine force that includes training in basic submarine operation, damage control, fire-fighting, flooding, watch-standing, warfare, escape from a sunken submarine, and specialty training (mine being electrical work). As mentioned earlier, I went through this twice, once as an enlisted crew member and once as an officer. But the most rigorous of the training a nuclear submariner goes through is Nuclear Power Training, a grueling year-long training that consists of six months of classroom study of subjects like math, physics, thermodynamics, metallurgy, and engineering, and another six months of practical training on and operating a real nuclear reactor. One might say I went through this training more than twice. When I completed the enlisted training, I was selected to be an instructor. So, I then had the opportunity to complete a two-year tour training other sailors in the operation and maintenance of a naval nuclear plant. Following that I was selected for commissioning as a line officer and then given the opportunity to go back through nuclear training at the college level at which officers participate. After retiring from the submarine force, I became an executive in a company that oversees the quality of operations of commercial nuclear power plants. During my civilian career, I became the Executive Director of the National Academy of Nuclear Training. In this role, I provided guidance and oversight of the training of all of the commercial nuclear operators, maintenance personnel, radiation control technicians, chemists, and engineers in the U.S. commercial nuclear industry. I also coordinated with other countries, such as Spain and Canada, as they developed their commercial nuclear training programs. I have seen the fruits of training, and I have seen what the lack of quality training yields. There are few things in the professional realm that I hold in greater respect than training.

What does all of this have to do with Nuclear Mustang Leadership? A lot. There are two types of organizations in the world: those that train and those that should train. Your training might be in a classroom, but it might not. It might be out in whatever facility you have. It might be formal, with lesson plans, curriculums, and administrative protocols and records, or it might be completely informal, with none of these. Whatever it is, when it is being performed, and how it is being performed will be key to the success of your organization.

So, what do you, as a leader, look for when you look at the training in your organization? There are two answers to this question. You look for the quality of the content, but also, and just as or even more important, for the effective reinforcement of good behavioral practices. These words are worth repeating because I doubt any professional teacher outside of the Navy's nuclear program would use them. The words are *good behavioral practices*. It has been my experience, time and again, that when training is ineffective in adjusting behaviors in the desired direction, these words capture the part of training most likely to be where the biggest problems lie.

Of course, the content of the training is important. This is where one learns that part A is to be inserted into part B, or for example in the training on softer skills, that the coacher sits down with the coachee in a private setting. In decades of observing the widest variety of training, I have found few instances of such content being incorrect. It might not be how I would do it, but to say the teaching was incorrect or wrong would in itself be incorrect or wrong. On the other hand, many, if not most, training activities do not adequately set and reinforce the behavioral standards expected within a company.

If you want to be an effective leader, observe any of the training going on in your organization and compare what you see of the Trainers, in terms of dress, speech, and behaviors, with what you expect of all your employees. Are they models? Are they dressed professionally? Do they treat each other, as well as the students, with respect? Do they use professional language? Do they follow procedures? Do they use error-reduction techniques? If you have certain expectations regarding the formality of face-to-face communications, are these expectations reinforced by the Trainers? If the answers to these kinds of questions are not an enthusiastic YES, then you would be correct in concluding your trainees are being trained; they are, however, being trained to do the wrong things. Don't let that happen. It is in training, or what I call the garden of training, that employees' knowledge and professional behavior will be nurtured, and they will either learn and practice what is expected of them or be led in the wrong direction. Good leaders know this, and they apply their attention accordingly. If you want good things to grow, you have to spend time in the garden.

# BUILD CONFIDENCE IN THOSE YOU LEAD

*"At its core, leadership is about instilling confidence in others. "*

-Howard Schultz

    Howard Schultz, an American businessman, author, and once CEO of Starbucks Coffee, hits the nail of the linkage between leadership and confidence-building on the proverbial head with the above quote. An effective leader knows this and knows as well that for one whom she leads to be the strongest contributor they can be; they must have confidence. Building that confidence is one of the leader's most important jobs. How do you build confidence in someone? I suggest the answer to this question again lies in the submarine community. Submarining builds confidence. Some in the civilian community might say too much confidence. But we'll leave that debate for another day, and focus instead on why and how it builds confidence, and more importantly, is there a downside to this confidence that we need to be alert to, and how can we best address that downside?

If you think about when and how you learned to ride a two-wheel bicycle, a picture would likely come to mind of you on your bike, with a parent running alongside and holding the back end of the bike so it wouldn't fall over. This would be after you had progressed from the training-wheel stage if you were fortunate enough to have had training wheels. At some point, when you were sure your parent was still holding on, you looked behind you, and that parent was off in the distance. It was then you realized you could keep this thing up and moving forward on your own. You probably were a little shaky on your next solo ride, but with each solo performance, you gained more and more confidence. Confidence in the evolution came when you saw your ability to do it on your own, to solve the problems you were faced with before you learned to stay upright on your bike. This is just a simple example of the point that confidence comes with successful problem-solving. On a submarine, one is always problem-solving, and the problems are solved successfully, or you are most likely dead. Suppose you're driving the submarine, a major hydraulic leak occurs, and the planes that control the sub's depth, by using the force of hydraulic lift as the sub "flies" through the water, go to full dive, pushing the sub toward the depth at which water pressure will crush the hull. This is a problem and not an unusual one. How do you solve it? Suppose one of the machines that removes carbon dioxide from the sub's atmosphere fails, and you know it takes all of these machines operating to maintain a reasonable atmosphere. What do you do? Suppose a second machine fails. I have experienced this on multiple occasions. Now, what do you do? Suppose a fire occurs on board, the sub fills with smoke, no one can breathe, and the sub is operating in unfriendly waters where it is not possible to surface. You can't call the fire department. What do you do? These and a myriad of other problems face submariners regularly. The submariner learns how to deal with each of them, first as a trainee, with someone holding the bike up. With each successful dealing, like the child who realizes that he is moving along alone on his bike, one develops confidence. Eventually, that confidence reaches a level that one believes they can deal with anything.

So, what's wrong with confidence? If you believe you can solve any problem, why do you even need anyone else? Therein lies the problem. The confidence leads to you cheating others out of the opportunity to try riding the bike on their own. If they don't get a chance to solve problems, then their confidence will not get a chance to grow, they will not learn to solve problems, and your crew will be a one-man show. Never a good idea.

What's the fix? Work through others. Hold the bike for them for a while, but eventually let them go. Stretch their capabilities. Let them solve problems, and not just simple ones. Critique their methods and results, provide constructive feedback, but let them do it. Hold the bike, but eventually, let them ride on their own.

# REMEMBER THAT IT'S ABOUT THE MISSION, NOT ABOUT YOU

*"A small body of determined spirits fired by an unquenchable faith in their mission can alter the course of human history."*

-Mahatma Gandhi

In my earliest days of submarining, when I was an enlisted crew member, a quite junior one, on a diesel-driven submarine, I learned one of the most important lessons one can learn in the Submarine Service, a lesson that also effectively conveys a key principle of Nuclear Mustang Leadership. I talk about this lesson in the opening pages of this book when I described my submarine experience of being underwater and having the compartment next to the compartment that I was in, called Maneuvering, begin to flood, and simultaneously discovering that the only means of escape was irreversibly blocked. I explained in that opening tale that no one other than me seemed bothered by the inability to escape as they calmly but quickly went about their responsibilities in the event of flooding. Their actions were effective. The flooding was stopped. The intruding sea water was returned to where it belonged, and we went about our submarining business.

Not long after this event, I learned, primarily through discussions with my peers, that one's actions on a submarine as a vessel of war are directed toward the successful completion of the mission, not to the well-being of any individual crew member. The quote of Mahatma Gandhi that opens this section gives a picture, and an accurate one in my experience, regarding how effective a real focus on a mission can be. On a submarine, if a problem occurs, and the boat is about to go to the bottom of the sea for the last time, every single crew member is acting to save the boat and continue the mission, not to save themselves. The circumstances of submarining are such that this has to be the case. If the submarine dies, all will die with it, and every single crew member's action may be important to not letting the submarine die. The mission becomes, save the sub, not yourself. As I learned this lesson, I was embarrassed at my thoughts in that flooding incident, my selfish thoughts about escaping. I never thought that way again.

"Casualties" like explosions, fires, and flooding were not unusual on the boats, at least on the ones on which I operated. They are prominent in my memory of submarine life more than any other aspect. It was these events that defined the submarine experience and further solidified my understanding of the principle described above whatever they are in, submariners are in it together. I am confident this feeling exists in many military groups, but what makes the submarine experience different is the highly technical and complex environment. Thousands of gauges, valves, and electrical components surround a person and are woven into a set of interrelated systems that can either save you or work against you or, in the worst case, take you unwillingly and unexpectedly to a final resting place at the bottom of the ocean. Every component must be understood and operated properly, sometimes in unorthodox ways. This environment is furthermore built inside a medicine-pill-shaped capsule called a submarine pressure hull and immersed in what is referred to as a "hostile environment," often frigid water, trying as hard as it can to put its incredible pressure on your body at any opportunity, to envelope you, and keep you with it, forever. On a submarine, the entire crew is relied on to operate and maintain every one of these components. Everyone must rely on everyone else, and that reliance is based on a well-founded and continuously building trust. All individuals' actions are geared not to save themselves but to save the ship and complete the mission.

Think about the above philosophy in terms of a business or any professional endeavor that involves the need for people to work together. The motivation of the sailors in the above scenario is the threat of dying. You either work together to focus on the mission and save the sub, or you die, along with everyone else. You can't use the threat of dying as a motivator in your civilian circumstances, but you can use the remarkable teamwork of a group of submarine sailors in the face of death as a benchmark against which to measure your own efforts to motivate your "crew." Do they even understand that you want them to put the mission of your organization first? Have you set that expectation? Have you motivated them - with compensation, recognition, and respect? Does everyone understand it's about the mission, not about themselves?

# LEARN TO RELATE TO ALL LEVELS AND THEN DO IT!

*"Leaders must be close enough to relate to others but far enough ahead to motivate them."*

-John Maxwell

As Mr. Maxwell, the American author, writer, and leadership expert, implies in his quote, it's important to relate to those you are leading. Remember that the only purpose of leading is to complete the mission, get the job done and done as well as it can be, and if you can't relate to those who will do the actual work, you will never get them to perform to their highest potential or do their necessary best. A real leader knows this and uses a three-step approach to establish or maintain that relationship walk, listen, and be seen and I mean really seen.

The first of these three has been touched on in several different sections of this book, the reason for the repetition being its importance walk. Walk in their shoes. Get out into the workspaces of whatever facility you are leading. Try doing what those you are leading are doing. Ask for their help to understand. They'll appreciate your effort. Once they conclude they can trust you, they will also tell you things you should know. This is the second of the three-step approach. Listen to those you are privileged to lead. And finally, the third step let them see you, not the organizational you, not the you with the manager uniform on, not the you who can and does affect their paycheck; but rather the real you, the person you are. If you do these three steps, you will be relating.

Many who consider themselves leaders will read this principle and think to themselves, of course, I know that, and I do that. Don't believe it. The operative word here is "relate". Many will feign relating. They will pretend to relate but not really understand those with whom they are trying to relate. They will not have had experiences similar to the experiences of those with whom they are interacting, will not have empathy for them, and will not connect with them. The would-be relaters often will even fool themselves into thinking they are relating. Do you really think a politician who takes off his tie and rolls up his sleeves at a county fair is relating to the farmers in his audience, the ones with the overly red necks, brought about by hours and hours, and days and years of hard work in a blazing sun? Do you think one who passes up her routinely eaten seafood delicacy or filet-mignon-caliber food to have a hot dog when she's trying to show that she's relating to someone at a fair is genuine? Can you honestly believe those she is mingling with won't see through this token effort?

So, how does one genuinely relate? If you have worked and lived "in the trenches" for any reasonable amount of time, as Mustangs have, you can already relate. You have been there; you have done that. But what if you have not? Then go find out what it is like to walk in their work shoes, and do your finding out when no one is looking. Don't turn your effort into a PR stunt. Bring a sandwich you made in a paper bag and sit down and talk to the people during lunch. You can talk about work issues, but it would be better to talk about them. How many children do they have? Their ages? What do their wives do? What do they most like to do in their spare time? Do you and they have any common likes or dislikes

sports? travel? Don't brag about yourself or
anything you have done. Expose yourself; open
up about your fears, shortcomings, and abilities
you wish you had. You don't have to be a
comedian, but if you can recall any humorous
events, bring them up, especially self-deprecating
events or anecdotes that show you're a real
person, not that much different from them at
heart. Don't react to anything they say. Just listen.
I once led a small team of highly intelligent,
technical people, all engineers and scientists.
They were all several levels below me in our
organization, and to a degree, their careers rested
in my hands during this assignment. In the course
of one evening dinner, I described an event in
which I had tried to buy a piece of cloth at a
Walmart. I think some were even surprised to
learn I shopped at Walmart, just like they did.
The cloth I was looking for I had intended to use
as a covering for a deer hunting blind that I had
built. The fact that I was a deer hunter, something
not likely to be included in one of my
presentations or speeches, also seemed to get
their attention. For some reason, as I described in
my tale, the sales lady, who was the typical,
friendly, down-to-earth country lady, mistakenly
concluded that I intended to make a dress for my

wife as a gift. She thought that was such a romantic thing to do and was so foreign in her world of macho outdoorsmen. I quickly picked up on her misunderstanding and played along, allowing her to believe that was indeed my intention. The story went on, and I'll spare you the details and the graphic country language I used, but it was quite an entertaining story. When the sales lady saw I had picked a dull olive-green color material, asking her if she thought my wife would like it, she somewhat hesitatingly agreed the wife was sure to love it. When she asked whether or not I needed such a large piece of cloth and I told her that my wife was a "good-sized woman," she stopped asking questions but continued to seem more than pleased at my apparent effort. The story went on until I confessed to the store lady the truth behind my purchase and suffered probably a lot less scornful looks and expressions than I deserved. But the important point of my tale of good-hearted deception is that when I had finished the story, one of the engineers, a young lady whom I had already identified as being an exceptionally intelligent engineer and of high potential, said to me, "Listening to that story I now have a completely different picture of who you are, and I appreciate that. I had taken you for some distant character who was all work without a humorous bone in your body. I now feel more comfortable when we're talking". Little did she know that she had just taught me a lesson. I had just been acting naturally, but she reinforced the value of that to the point that I adopted it as a routine part of my leadership tool kit. Honest communication is probably the biggest return you will get from relating to people. One more example: One of the more junior crew members on a submarine on which I served had made his announcement that he intended to not reenlist, and instead would leave the Submarine Force and the Navy at the earliest opportunity. His loss to the boat was a major one and came at a time when personnel retention was particularly low. Both the Skipper of our boat and the Executive Officer had asked him why he was leaving the service, and in his replies, both of which were sprinkled with a heavy, if not overdone, dose of "yes sirs" and "no sirs," he relayed no information that would help us address the problem of poor retention numbers. After these senior officers had their ineffective career discussions with the young submariner, I came across him one evening (of the many I worked) and had the opportunity to

have an uninterrupted discussion. It was an
impromptu meeting in his workspace, and I think
he sensed my genuine concern for him since I
was working clearly beyond normal hours, and I
still took the time to talk with him. I asked him
what he was going to do after he left the service.
His response, delivered in a direct but respectful
tone, believe it or not, was, "I'm going to blow
goats for a nickel a herd until I get my self-
respect back." Wow! How informative! This guy
just told me that whatever we had done, we had
not helped him to have any level of self-respect.
No wonder he was leaving us. What an
opportunity it was for me to continue a frank
discussion. He went on to tell me how he had
been so badly misled by his recruiter, who told
him that if he signed up as a "mess management
specialist" with the emphasis on specialist, he
would be like a chef and would work only with
officers. In reality, he was a waiter and
housekeeper for the Officers, whose wardroom is
sometimes referred to as the Officer's Mess. He
had nothing to do with cooking and did little
more than carry food from the tiny submarine
galley or kitchen to the officer's table in the
wardroom. Some of the officers treated him like a
waiter, giving him essentially no respect but

expecting exceptional courtesy and flawless performance from him in serving them, laundering their clothes, and cleaning their living spaces. In being honest with me he told me what the problem was. It was us. I wish this story had a happier ending, but it doesn't. The young man left the service. I just wish I had been able to let him know what a valuable contribution he made before his leaving.

If you don't know how to relate, learn how. If you're not comfortable relating, get comfortable. Most importantly, do it.

# ALWAYS PICK THE HARD STUFF TO DO

*"If you want to succeed, choose the hard path at every fork in the professional road."*

-WTS

My introduction to Nuclear Mustang Leadership came long before I had ever heard the term or would have even understood what it meant had I heard. The introduction, in itself, encompasses one of the many principles involved in such leadership, and that is, to be a leader of any type, one must first lead oneself in the direction of continuing improvement and advancement. A loser cannot be an effective leader. I, figuratively, stumbled upon this principle when, during my high school years, I learned to value the importance of forks. These are not the forks one finds on the table at mealtime but rather forks in the road of one's career.

I was not long out of high school when I faced one of those ever-important forks. I never considered my family to be poor as I was growing up, but with the wisdom born of hindsight, I can now see that we were poor. My father was a coal miner until the mines closed, and he changed his occupation to that of a manual laborer on a highway construction project. My mother was a factory worker who spent eight to ten hours a day standing on a concrete floor in a hot factory, ironing shirts that had just been made by some of her sewing colleagues as the shirts were prepared for packaging and subsequent sale. Our home was spartan, to say the least. I can still picture the first house of my life, a rented one with a total of three rooms, one being the kitchen and two on the floor above being bedrooms, one for my parents, and one for myself and my two siblings to share. We had electricity and running water, but only cold. Our toilet was an outhouse in the yard. I believe that this overly frugal beginning was one of several factors that played a role in forming the determination that would drive me for the rest of my life. At one point, after I had graduated high school and shown myself to be not ready for advanced education, I reveled in that first breath of real freedom that I suspect all new graduates experience. I could do anything I wanted to do. Based on my social and financial status, of course, my options were limited, but I did have choices. This was the first fork I recall facing. I could proceed down either of two paths, one easy and one hard. The easy path would have required the least effort, physical as well as mental. I could have chosen to lie around my parent's house and live off them as, unfortunately, some of my friends did. I could have traveled, to whatever degree my limited finances would have allowed, and just focused on having a life of fun and adventure with limited, if any, responsibilities. The lure of this branch of the fork was instant gratification. I could begin enjoying myself immediately no work, take it easy, no commitments, no responsibilities. The hard fork was completely different. It involved leaving my home of birth, essentially forever, setting off on my own, taking responsibility for supporting myself and expanding my essentially non-existent skill set. For reasons even I don't completely understand, in this first major decision of my professional life, I took the hard fork. I left home and enlisted in the Navy. This fork included the rigors of basic military training and the substantial and immediate loss of many of my freedoms. The marching band playing at the ceremony of boot camp graduation quickly faded as I stood facing the world in my new navy uniform with the one small, red stripe on my arm indicating I was a Fireman Recruit, an E-1, the lowest level of enlisted person in the U.S. Navy. At that point, before me appeared another fork in the road. Down the easy branch of the fork, I could see a shore duty, desk job assignment, or a fun assignment on a smaller ship or boat with a lot of physical activity but not much hard work, certainly not mental, and more fun and adventure than play. I might even see the world, compliments of Uncle Sam. But again, I chose the hard path of the fork. I volunteered for submarine duty and the associated hazards and life deprivations, but also the academic rigors of learning nuclear and submarine technology. My choice meant significantly reduced social life, long hours of academic labor, stressful physical and psychological testing, no small risk of personal danger, and regularly leaving home, friends, family, and even civilization for months at a time. But also, along this path lay the many opportunities to hone my technical and learning skills, opportunities that led me to this choice.

I soon learned from these experiences that a pattern of choosing the hard path at a fork in the road was a strategy for success. These forks are sometimes not obvious. Some don't even seem like decisions, but they are. For example, a highly successful friend of mine recently made a statement that applies well to me and could be looked at as one of those forks. He said, "I never felt ready for any job I was asked to fill. But I accepted each as an opportunity to broaden myself." I have felt similarly on the many occasions when I was either promoted to a higher-level position in the company I worked for or was assigned a task, civilian or military, that I knew was a stretch for me at my then level of expertise and knowledge. Accepting those new responsibilities, embracing them, and taking on those new assignments with gusto and enthusiasm was essentially a decision to go down a harder fork of the road than the one along which I could relax in the comfort of the experience I had already developed. I continued rigorously adhering to this philosophy of choosing the harder fork, and made a career in the navy, attended about twelve years of various education and training, obtained both undergraduate and graduate degrees in engineering, received a commission as a line officer, eligible for command of a ship, advanced to the rank of

Lieutenant Commander, and became the Chief Engineer on a state-of-the-art nuclear submarine. Many forks preceded and followed, and it was always the hard path I chose. I went on, after the Navy, to become an executive in a private company, attend the Harvard Graduate School of Business, retired a second time, and went on to start my own consulting business, sitting on several nuclear safety oversight boards of large electrical generation companies always ready to look for and choose those hard path forks.

It was the above experience that led me to recognize that I was not alone in choosing the harder path at the forks of life but that others who chose the same were the ones from whom I could learn and the ones I enjoyed associating with. The practice became, in my mind an attribute, an attribute of what some of my colleagues called a "hard charger". I also learned that everyone comes to forks on their life paths, but not everyone recognizes them as such, and many more choose the less difficult path. A good leader, however, looks for the forks, recognizes them, and doesn't just take the hard path occasionally or even some of the time. She picks the hard fork every time.

# HAVE AND SHOW HIGH ENERGY

*.  "High energy, you either have it or you don't. "*

-WTS

I discuss the personal trait of high energy in The Observant Eye, defining it by describing what it looks like. It looks like a person who moves quickly, covers a lot of ground, gets a lot done, never sees an unsolvable problem, volunteers for more, particularly so even when they already have a full plate of work, goes well out of their way to examine every aspect of and fully understand a problem; starts early; works late; swaps a dress suit for work clothes and gets out into the trenches with the workers to see what's going on, where a focus is needed, and how they can help the workforce. When a high-energy worker doggedly pursues every aspect of a problem or a task and ends up at a dead end, what do they do? They quickly come up with another approach and get right into it.

How does one get high energy? You can't. You either have it or you don't. That said, in the context of leadership, and particularly Nuclear Mustang Leadership, you can't just give up on someone you're attempting to lead just because they don't have high energy. You do everything you can to coach them on the various traits of a high-energy person, as discussed above and work to get that person to adopt and develop as many of those traits as can be adopted by that person. Seeking high-energy behaviors in someone is not unlike pursuing excellence. You never get there but are always striving to.

The concept of high energy, as it applies to work, can be a controversial one. I have even thought to myself at times that this is not a topic worth discussing. One either has high energy or not. People are born with it or not depending on a variety of factors, not the least of which is heredity. That said, I chose to discuss it here because I believe it is one of the most important traits a leader of people can have, and without it, or at least a reasonable emulation of the traits that define it, one can surely not be a good leader. Also, over the years, as I coached an uncountable number of people, I came to believe that although one might not be able to make someone high energy through coaching, one can coach people to exhibit the characteristics of someone having high energy and that, in itself, is valuable. For example, a high-energy person comes to work early and stays late. (I suspect with these words the controversy may begin to brew already.) A high-energy person doesn't waste time at work. Time wasters include overly long, non-work-related conversations, at the end of which the state of the job is the same as it was when the conversation started because they have added little to no value to the work. There are numerous other time-wasters, many of which involve technology, and especially social media.

The internet is replete with advice regarding high energy, much of it esoteric and unrelated to job performance. But even that advice related to job performance often gets into other non-job-related areas like maintaining a healthy diet and getting enough rest. I'll leave the self-care advice to others and will provide a short discussion on each of what I see as the key attributes of a high-energy employee.

To expand on the concept of high energy and show how it applies specifically to getting things done through people, I will start with the core principle that an effective leader has high energy.

I have been blessed, some might say cursed, with high energy.

Some might say that having high energy equates to working hard. That is incorrect. More on working hard a bit later, but the universe of high energy characteristics includes working hard. It also includes working smart, having the correct attitude and the proper motivation, and, most importantly, working effectively. One cannot work effectively without having the ability to correctly prioritize. The ability to effectively prioritize is likely one of the most important skills one must have to be a good leader. It is also the skill that, by its absence, causes many of those in charge to fail.

Much has been written about work/life balance. If you have read some of the literature on this subject, forget it. There is no such thing as work/life balance. But there is prioritization, and it cannot be general prioritization. If you put the highest priority on your personal life and family, you likely will not do well in your job. You wouldn't if I were your boss. If you put the highest priority on work, you likely will lose your family. You would if I were your spouse. The prioritization has to be applied to each specific situation. If your young child is starring in her first school play, and you put the highest priority on staying at work and checking your e-mail or in-basket, as you would if you generally placed priority on work over life, you would be making a mistake. The best leaders don't make these kinds of mistakes. If, for another example, you choose to take several days off to take the family on a long-delayed beach vacation during a time when the company is struggling with a major setback, you would also be making a mistake. Once again, good leaders don't make these kinds of mistakes. A good leader weighs each situation on its own merits, with a bias toward support of the company, from which comes the ability to support one's family.

I share my thoughts on the personal characteristic of high energy in *The Observant Eye,* and, in addition to explaining what it is, I recommend therein never hiring someone who doesn't have high energy. I will repeat little here that I have written there, other than to emphasize those points that are most important. I will instead expand on the concept and how it applies specifically to getting things done through people.

I will put my humility aside for a bit and provide several excerpts from some of my past performance evaluations to show by example what high energy looks like from the perspective of those above me in various organizations.

➢ "(In his role as a Naval Nuclear Power Instructor) he has spent whatever extra time was necessary to show slow-learners the way to success."

➢ "(In his effort to advance) he has studied courses in English composition, history, and mathematics in his spare time."

➢ "He continually strives to improve himself by correspondence courses and self-study."

➢ "He has a zeal to excel in every undertaking."

➢ "He set the pace for other officers (in submarine qualification) by qualifying early with excellent knowledge."

➢ "He is a very methodical worker who has tireless energy, being able to work many hours without sleep."

➢ "Although assigned approximately twice the administrative workload of any other division officer, he consistently produced high-quality work on or before the required submission date."

➤ "When a deficiency was found in a vital piece of submarine equipment that would adversely affect the submarine's schedule, a shipyard estimated several days for repair. He volunteered the ship's force to do the work, himself to lead it, and completed the job in less than three hours."

➤ "He does everything superbly as the Chief Engineer of a nuclear-powered fleet ballistic missile submarine undergoing extensive overhaul in a civilian shipyard. This job has often proved too much for many officers to handle. He not only handles the job but excels in all aspects of it."

The world is not short of examples to follow in the trait of high energy. I will close out this topic by describing just one of many examples, which comes in the form of Adena T. Friedman, the CEO of NASDAQ, the global electronic marketplace for buying and selling securities. She has also been listed repeatedly on Forbes's annual list of the world's most powerful women. And she started as an intern! In a Time magazine leadership forum, Ms. Friedman was asked how, starting as an intern at NASDAQ, she was able to differentiate herself from others. She responded, "I consistently asked to do more. I definitely was not sitting there waiting for someone to tell me what to do next. I found that if I had some free time, I'd go over to someone and say, "Do you have anything I can help you with?" I was always willing to take on more. That is a picture of high energy.

# WORK HARD

*"Working hard provides a competitive edge over peers that awaits anyone who chooses to use it."*

-WTS

As stated above, the universe of high energy characteristics includes working hard. What does that mean? It means putting every ounce of energy one has into whatever effort they are working on. It means starting work early and leaving work late if required to get the job done, not to the best of one's ability, but better than that. It means getting the job done as best as it can be done. It includes an understanding that the" best of one's ability" can always be made better. If writing a paper, it means writing the paper in draft as many times as it takes to make it as good, as clear, and as concise as possible. I discovered the value of working hard early in the first of my careers, that was operating a diesel-powered submarine in the U.S. Submarine Force. My first job, as mentioned earlier, in simple terms, was a dishwasher. In reality, I was in that position I said was, at the time, called mess cook, the most junior person on the boat, as we call submarines. Among my responsibilities of washing the dishes after each meal, cleaning the galley or cooking and eating space, and taking out the garbage was completing my qualification on submarines. Qualification included having demonstrated the ability to fully and effectively carry out the responsibilities I might be assigned to, such as those of a lookout when on the surface and those of operating the planes that control submarine depth when submerged. It included knowing the configuration and capacity of essentially every tank and pump on the boat, the capacity and configuration of essentially all of the electrical and propulsion components, and the function and construction of every major valve. Submarine qualification required a tremendous amount of reading and study which was then followed by "checkouts" or tests administered by more experienced crew members to verify one's knowledge and ability. I approached this qualification effort with a vengeance. I rarely slept. When other not-yet-qualified individuals were going ashore to take a break, relax, or maybe watch a movie on the boat in the evenings, I was working on my qualifications. I completed those qualifications earlier than anybody else at my level and was awarded the much-prized silver dolphins of the submarine force. It was that experience that opened my eyes to the fact that through hard work, much can be accomplished. Working hard became who I was. This approach saw me move up the enlisted ranks, attend, and then, as mentioned earlier, be selected for an instructor's role at one of the highly acclaimed nuclear power training units of the U.S. Navy. It further saw me be selected for a commissioned officer program and then eventually achieve the rank of Lieutenant Commander and be the Chief Engineer of a nuclear-powered submarine. The model continued to serve me well when I retired from the military and went to work in a prestigious nuclear oversight company in the commercial nuclear power world, where, again, I quickly moved up the management chain, being hired as an entry-level employee and retiring as a vice president and officer of the company. Following my second career, I went on to form my own consulting business and again did extremely well.

If I agreed to do a job, any job, I put in more than 100 percent effort on it. If it was worth doing, I thought, it was worth doing as best I could do it, and my best was something I was always trying to improve. When I retired from my second career, I had 39 vacation days on the books, and I had taken no sick days for several years. I never waited to be put in charge; I took charge. Again, that approach of working hard has served me exceptionally well.

Be aware that some disagree with my point above about working hard. My friends at the Harvard Business School had a headline in their August 2022 edition of the Harvard Gazette. It read: "When parents tell kids to "Work Hard," do they send the wrong message? They went on to say, "If you are learning that effort is the way to achieve success, and (I'd like to point out this is a BIG "and") you see people who have less, you might assume they didn't work hard enough as opposed to recognizing the systems and institutions we know can stand in the way." To my friends at Harvard, I say I agree. Don't assume anything about others but work hard if you want to get ahead. Work hard to set an example for those you lead. Work hard if you want to be a good leader.

# FOCUS ON THE TASK NOT THE CONSEQUENCES

*"Instead of focusing on the circumstances that you cannot change focus strongly and powerfully on the circumstances that you can."*

-Joy Page

Think about the following situation: You're driving a car during a wintry day, and the car goes into a slide on an icy road surface. You know what to do - turn the wheel in the direction of the slide, and what not to do - slam your foot on the brake. You also know that if the car continues to slide, there are potential consequences. You will likely either hit the large bank on either side of the car or hit another car, either of which will likely result in some extensive personal injuries. If you concentrate on your actions and your indications, you likely will do as good a job as you are capable of doing. If you concentrate on the consequences, you will only detract from your mental effort to do the right thing.

Joy Page is a highly successful American actress who unfortunately died some years ago, but not before providing the introductory quote for this section. It is unlikely she was thinking about submarines when she gave birth to the above quote. Her words are, however, a good way of expressing a fundamental principle of submarining.

Tie the thought of Ms. Page's quote in with that of my opening words for this section, and a key principle of Nuclear Mustang Leadership begins to surface. I have never slid down an icy road in a submarine, but I have had a torpedo that we had just fired, one heavily laden with explosives, turn around and come directly back at our submarine. Every crew member did exactly what they had to do. The boat was quickly rigged for the explosion that might occur; communications between all parts of the submarine were instantly established and then maintained; crew members were positioned and ready to combat the flooding that would occur should we take the hit. And the submarine was quickly and properly maneuvered, in accordance with our training, to avoid the errant torpedo, or "fish" as we called them. A catastrophe was avoided. It might not have been had the crew, instead of focusing on the actions needed, over which they had control, let their thoughts and focus move onto the consequences of the torpedo ending our day. As another real-life example, when I was the Chief Engineer on a nuclear submarine, for reasons too technical and off the point to be worth describing here, we had lost all power on the sub. I was not in the spaces where the engineering equipment related to the loss of power occurred, but one of my most experienced enlisted men was. Ordinarily, he would know exactly what to do and how to do it. For reasons no one will ever know, he let the seriousness of the situation play with his mind. Were we not able to relatively quickly restore power, among other problems, the nuclear core could overheat and possibly even melt, causing indescribable damage. This gent had apparently begun to worry about the consequences of not keeping the reactor core cooled. This led him to become gripped by fear that in turn, led to panic, rendering him incapable of performing the actions needed to recover from the power loss. Being in a different part of the sub at the time, I had received indications of the problem and quickly went to the area where the needed actions would be performed, and where this petty officer now stood, essentially ineffective in every way. What happened next is an element of a different story later in this book, but suffice for now to say that something happened to get this petty officer's mind off the consequences just long enough to snap back into his normally highly effective self, focus on the actions needed, and recover from the situation.

I remember the cases above well, and I remember many other similar cases, in which a major casualty was avoided by the entire crew being focused on what they had to do and not getting distracted by the potential consequences. It is not that the crew members were brave, although they were, but rather that they were aware, because of their many experiences, that thoughts about consequences are not generated in an effectively used mind.

Not only is focus important but what the focus is on is what is really important. Face each problem as it occurs, and do your best to address the problem without being distracted by the consequences if you don't address it. Letting the mind wander into the realm of consequences does nothing more than detract from the concentration needed to perform reactionary tasks to the best of one's ability. It is also important to note that what allows one to have the confidence to focus on the task at hand and not think about consequences is training and practice, more training and practice, and more training and practice. The consequences of not doing things right then don't even come into consideration. If things are performed properly, things will turn out well. If not performed properly, thinking about the consequences in advance will be of little help anyway. It is amazing what can be accomplished with a concentrated focus on the right things. The Nuclear Mustang Leadership aspect of this discussion is, of course, once again about setting an example yourself by not focusing on consequences and setting expectations of those reporting to you to do similarly.

# TAKE NOTES AND BE A SELF–STARTER

*"Everything cometh to he who waiteth as long as he worketh his ass off while he waiteth."*

-WTS

A good leader is not a collector of information. A good leader is a hoarder of information, an extremist at collecting information that is performance-related. Such a leader keeps notes on everything performance-related. For example, I have notes of just about every intetellrview I ever conducted over decades of management oversight interactions, with the names and individual people and company identifiers removed to protect their privacy, of course. I have my military fitness reports from my many-years-ago career in the U.S. Submarine Force. I have the feedback sheets that I received as a civilian member of teams that assessed the performance of highly technical commercial facilities and their management. My notes go back more than fifty years. Why do I do this? To improve. I use these notes. I periodically revisit them when preparing talks, when thinking about problem solutions, and when looking for self-improvement ideas. The notes are like data in an advanced degree program, and I continually analyze my data, looking for new insights.

I keep notes, not only on myself, but on everything, and I analyze them looking for improvement opportunities. I do a lot of backcountry adventure-type fun activities, like wilderness trout fishing in places like Montana and Idaho, and after every trip, I write a trip report. Do I have to? Of course not. But I do. And before I go on my next trip I'll go back and review related reports of past trips to see how I can make this trip better. Is this hard? You bet it is. There are a lot of things I'd rather be doing than sitting down and writing up detailed notes and trip reports, but I don't. Why? Because I'm not one to sit around and believe everything will come to me if I waiteth. If you want to try a useful exercise, start a detailed, and I mean detailed, recording of your interactions with the medical community doctors, physician's assistants, pharmacists, and anything related to your health. Periodically review it, and I can just about guarantee you will find many useful informational tidbits, as I have, with the potential to directly affect your health. You will likely find that you have not been doing everything that various physicians recommended you do. You will also likely find that physicians are good, but they are not always consistent with each other in their guidance. Are there opportunities for improvement in this exercise? Bet on it.

I learned the value of note-taking during my time on submarines, and only later did I realize how valuable this practice is if one is interested in doing a good job and advancing. Ensuring the equipment on a submarine operates properly is important, especially if you have any desire to live a long life. This equipment might be the atmosphere scrubbers that remove carbon dioxide from the air you breathe. It might be the oxygen generators that put out that ever-pleasant stuff your lungs scream for when they don't have any. It might be the ice cream machine without which, the crew would tell you, they could not survive a several-month patrol underwater. I have maintained or been responsible for others who maintained just about every piece of equipment on a submarine. Can you guess what is the one item most important to the long-term proper operation of this equipment? The notes people have kept on it. In the Navy, we called it machinery history. Every bit of work and every shortcoming of a machine is recorded. When that machine exhibits a problem, say a higher-than-normal operating temperature, the first place one looks to understand the issue is in its machinery history. Oftentimes the history will provide insight into the needs of the machine, information that will help that machine run better. My notes on myself are my personal machinery history.

My intention here is to not only encourage would-be leaders to take notes on themselves and to be hoarders of information provided by others but rather to also share what I found during one of my self-history reviews in preparation for this book. This note-taking effort is not unique to submarine life. I have used it in a variety of personal as well as occupational endeavors. I have been very fortunate (the kind of good fortune that seems to come with hard work) to have done well in every professional endeavor in which I have engaged, in each of my multiple careers. In reviewing what positive feedback I had received over a period of about eighteen years in a civilian company, one particular description among my strengths that was frequently mentioned was "self-starter". I honestly had not thought about this before this review and it was not until I saw the pattern in the notes that it became apparent. And when I thought about it, I was and still am a self-starter. I wait for no one. If I see a job to be done. I do it. If I see a problem, I work to solve it. I never wait for someone to tell me to do these things. And I don't think about being a self-starter. I think about being a good leader and that's just the way it is in the world of leaders. Why wouldn't you be a self-starter? Why would you want to wait until someone told you to do something? Why would you wait to do something later when you could do it now? Another quote, similar to but different from the one opening this section is one of which I cannot recall the origin, but it also is one of my favorites. "The only thing that cometh to he who waiteth is that which is left behind by he who didn't waiteth." I wait for no one. A good leader waits for no one. Be a self-starter.

# CONVEY CREDIBILITY

*"Credibility, like virginity, can only be lost once
and never recovered."*

-Charley Reese

As an enlisted sailor in the middle stages of my military career, my uniform reflected credibility. I had one hashmark that indicated four years in the service. But at the same time, I wore the "crow" of an E-6 or Petty Officer 1st Class. This combination, which said I had advanced significantly in a relatively short time, almost shouted credibility as a squared-away sailor. My early years as a commissioned line officer in the Navy also reflected credibility. I had earned my silver dolphin submarine pin by qualifying in submarines as an enlisted crew member. I wore the pin proudly, even though I had converted to the officer ranks and uniform. The silver dolphins advertised that I had previously been enlisted and enlisted for at least long enough, ten years in my case, to become a qualified submariner. The experience shown by this combination of pin and uniform gave me instant credibility in many circles. I share this information only because it supports my credibility when talking about the importance of credibility.

How do you get credibility given you don't have the opportunity to travel the unusual path I did? It isn't hard to describe what you have to do. It is, however, hard to do it. The recipe is basic, and it has three components: Successful performance; in-the-trenches experience; and truthfulness. If you have these you will have credibility within the scope covered by these attributes. You can advertise your credibility, but you don't have to. It will display and advertise itself. That said, if giving a presentation or in some other short episode interaction, briefly flashing one's credibility in the opening can be useful.

As for the three elements of credibility in my view, successful performance shouldn't need much discussion. It is a statement of common sense. If you want to have credibility in a field, you must do well in that field. If you are not doing well, fix that before worrying about credibility. The second element, in-the-trenches experience, is one that I may well be alone in including in the essential elements of credibility.

The reason why I say I am likely alone in considering the importance of this element, and the reason that some will resist this belief, is that many don't have that experience and are not willing to exert the effort needed to get it. Examples of its importance are aplenty in *The Observant Eye,* so I'll leave the reader to refer to that book rather than repeat the content here. If one is not fortunate enough to have had in-the-trenches experience, all is not lost. There is an alternative. Make a concerted effort to experience, and I mean really experience, what those in the trenches experience regularly. By "really experience" I mean getting with the people who do the real work. You can't relate to conductors on a train by riding a train. Do what these people do for a while. Check some tickets; interact with some difficult passengers; supervise the loading or unloading of some cargo. Admit that you don't know everything, particularly the skills they have. Be humble. Show them the respect they deserve. Ask them questions about their job. Use questions that show an appreciation for what they do and express gratitude for their showing you how it's done. There's a measure for checking the level of your success in one of these efforts. If you don't come away from the experience tired and dirty, you probably didn't engage enough.

The third element of credibility may well be the most important of the three, truthfulness. In other terms, don't try to bullshit people, or try to pretend you are somebody who you are not. I consulted for several years in the healthcare community and was quite successful in helping them improve their patient safety performance. Usually, the first thing I did when interacting with the professionals of that community, for whom I have the highest regard, was tell them I am not a medical doctor and know no more about the practice of medicine than any reasonably well-read layperson would know. I then would go on to describe my experience in situations and scenarios in which I was confident they sometimes often found themselves. By exhibiting humility and letting the information I provided speak for me, rather than presenting as a know-it-all braggart, I was able to gain their attention, and then their respect. From that point, we usually moved on to a good and useful exchange of helpful information. And finally, as Charley Reese, that American syndicated columnist, so wisely said at the start of this section, one must protect her credibility, because it can never be recovered. How do you protect credibility? Based on my three-element definition, the only variable element is truthfulness. Be truthful in ALL situations and you will protect your credibility.

# MONITOR/OBSERVE

*"All of us are watchers of television, of time clocks, of traffic on the freeway but few are observers. Everyone is looking, not many are seeing."*

-Peter M. Leschak

One piece of advice I would offer any aspiring leader before getting into the details of how to proceed on the path to being such a leader is to look where you're going. Sounds simple, doesn't it? It's not. I wrote a book on the subject. The book, which I have already mentioned multiple times, has sold thousands of copies and has been adopted by several large companies for use in training their personnel. I once again encourage anyone holding this book to put it down and go first read *The Observant Eye Using it to Understand and Improve Performance.* I am striving to avoid repeating much of its information here. However, as a capsule summary, suffice to say that observing is more than watching. The opening quote by author Peter M. Leschak, author and wildland firefighter, captures the point well. It is watching and mentally comparing what one sees with what that activity or condition would look like if it were at the level of excellent performance.

A real leader knows the gaps to excellence in her organization because she has observed them. And, she pursues closing them constantly, monitoring to ensure progress. Of course, she listens to what people tell her, and she reads reports, but to truly understand where the state of progress is, she observes, looking for results, not just efforts or actions. For example, I once consulted with an organization whose work included handling extremely dangerous high explosives. They were quite pleased with their progress because of the establishment of an observation program that required managers and supervisors to get into the workplace and identify performance issues. As is often the case, establishing a "program" gives one the feeling of progress, of accomplishment, and the self-satisfaction that sometimes comes with that sense of accomplishment deludes one into thinking real progress has been made, even though results-oriented progress might never have occurred. When I probed to better understand this organization's sense of accomplishment, I found that no one knew of any metric or measure of their organization's performance, such as how frequently their safety barriers had been violated, how many other safety infractions they had, how many errors they had observed during chemical analysis, or during the control of radioactive materials, or even how many events they had or what the trend was for the number of these events. The bottom line is they had no idea if they had progressed in results or not. At the time I asked, I was aware that a number of these infractions had occurred during the short time I had been interacting with this organization. My interest in these numbers was based on strong evidence that the frequency and trend of minor infractions can often be a forerunner of events of greater significance

The preceding advice deals with pursuing problem areas, and that is appropriate; however, one other piece of advice that took me years to learn is to be aware of the absence of problems in an area. Just about any area one can think of has some type of control in place, complex controls for complex areas and simple or basic controls for straightforward areas. For example, in a surgical area, one basic control in place to ensure that surgery is performed on the right body part of the right patient is to mark the site of the intended surgery in advance of the operation. If work is to be done on a person's right wrist, an "X" or other annotation is marked on their right wrist. If I were in charge of such a facility and knew confidently that we never have problems with surgery being performed on the wrong body part, I would pay particular attention to assessing performance in this area. Why? Because, and you might want to write this important tip down, lack of problems breeds complacency; complacency leads to weak implementation of controls in that area, and weak controls usually lead to problems.

If you are or aspire to be a leader of others, a good exercise well worth the time it takes to execute is to ask yourself, what is my radar? What do I look at to understand what is going on in my organization? Then, consider how this monitoring effort stacks up with current performance issues in your organization. If I were a betting man, I would wager that if doing observations is not on your list, you'll likely conclude it should be. I would also encourage keeping the following basic observation techniques at the forefront of everyone's minds:

- Think performance-based and focus on IMPLEMENTATION rather than just the existence of procedures. Compliance is good but not enough; rules don't cover everything.
- Expect consistent adherence to expectations
- Ask so what? The answer to this simple question often adds a notable amount of insight to the topic being discussed.
- Pull the string, meaning that if a seemingly small problem becomes apparent, don't dismiss it as a small problem but rather continue to pursue the issue to see the depth and breadth of it.
- Look beyond the activity being observed: Are people trained, qualified, and supervised; Are processes clear, easy to follow, and efficient; Are surroundings clear, clean, and well-lit?

A good leader not only observes and expects those working for her to observe as well but also encourages all elements of organizations to engage in observation activities. My years of experience in this field have shown me that getting an organization to do its own observations of activities has many benefits:

- Fosters continuing improvement
- Changes how people look at things
- Empowers the staff to assist the organization in improving
- Moves the organization to a stronger safety culture

I'll close this section with one more example of the usefulness of observation. As a consultant, I was once asked to visit a U.S. government facility that had, to the best of my knowledge, never had an "outside" person come in and look at their activities. I walked through the facility and simply observed the activities that were in progress at the time. I then provided the head of the facility with a short writeup of what I saw, which included:

1. A high-risk evolution governed by a detailed procedure that contained numerous notes and cautions, was performed by each of the three stations involved in the procedure setting the procedure aside and each using a set of notes handwritten on a small card. The card consisted of an abbreviated checklist of the steps to be performed. They lacked any of the notes and cautions. And for each valve or switch that was mentioned, the notes gave the

position but failed to mention if the component needed to be positioned or was to already be in that position.

2. Communication between the various parties involved was poor at best. Responses to directions were often, "Roger that". When one party was directed to verify no leakage from a mechanical component, the response was, "It's clear". Much of the communication in what was a high-noise area, was done with hand motions, such as putting a thumb up in the air; twisting a wrist, or holding two fingers up in the air. Three managers observing these activities found all of these practices acceptable.

I was told by the Plant Manager after I made my report that my observations "rocked the organization". What I saw was apparently the way business was done at this facility, and none of these behaviors met senior management expectations. There was a complete disconnect between senior management and the workers.

Enough said about observing. It can be invaluable. Do it, do it well, and expect others to do it also. Make observing a way of life if you want to be an effective leader.

# BE AWARE OF DENIAL

*"That which is denied cannot be healed."*

-Brennan Manning

I have included herein a section on denial and its dangers because I have seen it on so many occasions in organizations that lacked effective leadership and had notable performance problems. And in all cases, the denial was a barrier to improvement. I will again touch on the state of denial later, in the section on dealing with criticism, which often seems to be a fertilizer for the growth of denial. I could find no better example of a person who is the opposite of one in denial than Brennan Manning, an American author, laicized priest, and public speaker. He was also, at least until 2013, when he died, the kind of person well-qualified to talk to people about how to accept their problems and face them head-on.

It is interesting to note that I cannot recall a single instance of denial during my two decades involved with submarines, and there's a point to be made here. Success or failure on a submarine means success or failure for the entire crew. In simple terms, success is living and failure is dying. If the boat dies, everyone dies with it. What's the point? If those you are leading are truly focused on a common mission, and if they understand that nothing less than total teamwork is expected of everyone, then they, like the submarine crew, will not tolerate denial among any of the individuals or groups. Such total dedication to the mission is not often found in the commercial environment, and it is the job of a leader to not only avoid getting caught up in a mode of denial but also to be alert for and not tolerate any person or group getting caught up in that mode.

I have been especially fortunate to have spent my second and third careers working in the commercial nuclear power industry, where the employees are particularly intelligent and motivated to do a good job. This kind of people is focused on consistently good performance because doing a good job is synonymous with sound and safe operation of the nuclear facility and that leads to the highest levels of safety, affecting not only themselves, but also their friends, neighbors, and families who often live nearby the nuclear plant. All that said, it was not that unusual to find people inadvertently caught up in a mode of denial.

I once consulted with an organization that had received several reports from external agencies pointing out performance shortfalls. The scent of denial was evident in the fact that all of the reports of this type had been produced by outside organizations. Internal reports painted an amazingly different picture. Of course, to understand the issues, I spent most of my consulting time, not talking to managers in their offices but wandering throughout the power plant, walking in the shoes of the workers to the best of my abilities.

As I interacted with the workforce, I soon found a sense of denial, evident in the following. Whenever I asked about shortfalls in one of the elements of the organization, people of that element were quick to point out improper behaviors by those in other units that were impacting them. For example, the people in the security wing of the facility, with responsibilities including those for controlling explosives and weapons firing activity, said that people from other organizations violating their safety barriers was their biggest issue. Pointing to the problems of others when asked about your own is usually indicative of an organization that thinks they are performing at a level considerably higher than they are. What they are doing is hoping they are performing well, and as some wise person once said, hope is the denial of reality.

Personnel in the healthcare industry, where I have also consulted and spent considerable time in the work trenches, interacting with the workers, regularly display a sense of denial. Hundreds of thousands of people die each year from preventable medical errors, yet the internet is replete with comments from caregivers that the problem is not with them but rather with "the system." I have yet to understand how "the system" holds the saw or the scalpel when the leg is amputated from the poor lady who came into the hospital because of a problem with her digestive system.

Denial. A real leader will be alert for it. Don't tolerate it, even within yourself. A leader continuously works on improving herself, and a key part of this effort is accepting criticism well. Even if you believe the criticism is unfair or

inaccurate, recognize that it provides an
opportunity to get better. Learn from it.

# MANAGE CHANGE

*"Changes are inevitable and not always controllable. What can be controlled is how you manage, react to, and work through the change process".*

-Kelly A. Morgan

One of my nuclear colleagues once told me, "S--- happens. Change should not. Manage it." I mentally filed this with the many important pieces of advice I have received over the years.

Real leaders know that things change. This has been recognized since the time of our founding fathers when Benjamin Franklin said "When we are finished changing, we are finished." For example, shopping malls are withering under the onslaught of online shopping. When was the last time you saw a landline phone? The first submarine on which I served, had to come up to on or near the surface to get fresh air every few hours. We also had to refuel the diesel engines that powered the sub every week or so. But, like all things, subs change. The last submarine on which I served typically stayed deeply submerged for months at a time, and the newest submarines are being designed to never have to refuel for the life of the submarine.

Experienced leaders also know that change can be hard. Sea-going colleagues on my first diesel-driven submarine refused to accept that diesel boats were on their way out and that extensive new training was required to operate the highly sophisticated nuclear-powered submarines. The call of the diesel boat crew, in defiance of the inevitable, especially when drinking in the bars near the submarine piers, was "diesel boats forever". But subs, like all things, change, and despite those calls, the last diesel-powered submarine, the USS Blueback (SS-581) was taken out of service almost 30 years ago.

Like diesel boats, horseshoes are also gone, victims of change. In the 1980s I worked for a large electrical generating company that, at the time, still had Blacksmiths on the payroll in the 1980s! These metal workers had at one time been used to service the shoes of the horses that pulled service wagons around the city. A horse hadn't been seen around the company for decades. This is just more evidence that change can be hard, and resistance to it can be great.

Like many elements of business, there are as many offerings of solutions to problems as there are problems, and the "change model" is the offering for the problem of dealing with change. Just go to your computer, search for "change model" and take your pick. But Kelly A. Morgan, an Inspirational Speaker, Author, and Business Coach, captures the basic point about change in our opening quote of this section. We need, however, to get beyond the basic point. A good leader is wary of the offerings of easy fixes, gimmicks, and the trappings of help that are, in reality, little more than process-focused efforts that take time and money and produce little other than more work for the staff and leaders. I have seen more loss than gain as a result of aggressively employing "change management models". The typical problem is that more energy goes into the paperwork and bureaucracy of a change management initiative than into managing the change. It is truly ironic that one of the bigger changes in some companies, and one that is often not managed well, is the additional effort and resources needed to implement a change management initiative.

The problems described above are not unique to change management initiatives. There is a broader point of interest here. It is the nature of most material items, and even more so of initiatives, that they consist of both a core group of characteristics and a plethora of add-ons, which I refer to as "frills". The core group is what makes the item or the effort what it is. Consider, as an example of, a material item, a rental house on the beach. Core elements of such an item that make it what it is would be: it must be a house; it must be for rent; and it must be on a beach. If it has these core elements, it is a rental house on a beach. Someone may own a similar facility that is not a rental house on the beach but wishes to rent it as a house on the beach. It might be a mobile home, an apartment, a condo, or maybe even a shack. It might be several blocks from, but not on, the beach. It might not even be close to the beach. Since core characteristics are not things that can be easily changed, to facilitate rental, the owner may try to compensate by adding frills that might be attractive, attention-getting, or even desired, but that distract from the fact that some core characteristics are missing, and it really is not a rental house on the beach. They might offer numerous beach toys; maybe a cart that can be used to transport gear to the beach; maybe a cute little tiki bar that they've built out in the backyard, or an open fireplace, or a grill. They might add a swimming pool. Look through the frills and you'll see the bottom line - this is not a rental house on the beach. Did you ever think a change management effort could be looked at like a rental house on the beach? It can. The frills of the effort would be a change management consultant; a specially designated change management team; a change management procedure; and special change management forms of one sort or another. But a real leader will tell you that the core elements of effective change management are those listed below. If it doesn't have these, no matter how many frills it has, it is not an effective change management effort. Said another way, it is not a rental house on the beach.

Core elements of an effective change management effort:

- Recognition
- Direction
- Communication
- Motivation
- Support
- Persistence

First, and most importantly, recognize that you are going to make a change, that it will impact your organization, and that it needs to be managed. Don't just let change happen. If one drifts along every day, distracted by the challenges that face every worthwhile endeavor, change will happen. It will happen TO you. Don't let it. Manage the change, and whether that change turns out with positive or negative results will depend on you and how well you manage it.

Set the direction in which you want to move. Define a goal for the change. What do you want to achieve by the change?

Take the advice offered by Peter F. Gallagher, an author and change management global thought leader, and communicate. To use his words, **"Effective change communication is** at the heart of successful change, it acts like the blood in our bodies, but instead of supplying vital oxygen and nutrients, communication supplies information and motivation to the impacted stakeholders". So, communicate until you are convinced you have over-communicated, and then communicate a lot more. Tell people the change is coming, how it will be accomplished, what it will entail, how it will affect them, and what specific role they will play in it. When the change is in progress, keep them updated. Continually remind them why the change is important to them.

Motivate people to support the change. Make motivation a part of every communication. Apply consequences, especially to reward those who actively support it. Identify those who are most supportive and use them to add to the change momentum. Nothing helps motivation more than making people a part of the change so they feel ownership for it.

Support the change. Provide the resources needed to make it. If you don't control the resources, infect the one who does with your passion to support the change. If the change involves changing behaviors, model the desired behaviors yourself. Publicly recognize and reward those behaviors in others.

And finally, persist in moving forward. Change can be hard; resistance to it is inevitable and can be strong. Some will tell you the change cannot be made. Don't believe them. Accept no advice that doesn't move the change forward. People impacted by the change will want to see the confidence of their leader. Show that confidence. Don't just let the change happen. Manage it.

# REMEMBER TO GIVE & TAKE (CREDIT & RESPONSIBILITY)

*"Leadership consists of nothing but taking responsibility for everything that goes wrong and giving your subordinates credit for everything that goes well."*

-Dwight D Eisenhower

I'm not sure that I would agree with the impressive General Eisenhower that leadership consists of "nothing" but what he states, but his quote makes a point that is proven to be accurate over and over. Activities in any universe can be simplistically categorized into those things that go well and those things that do not go well. Let's talk about the things that do not go well first. Any organization geared to moving forward will be interested in these because whatever it is that leads to this result has to be corrected or changed to achieve progress. One can do exhaustive investigations into these cases to identify underlying causes, but a better approach is to have everyone examine their own consciences to determine what piece of this pie is theirs. Following this line of thinking, you, as a leader, are responsible for everything that happens, so that should be the first piece of information on the table "It is my fault." By taking this approach, you will accomplish two things: first, you will set an example for others to follow; second, you will be admitting a truth, because whatever happens that does not go well is your fault. You might not have provided enough resources, enough direction, enough training, enough guidance, or enough of any number of supportive elements. And don't just accept responsibility perfunctorily, mean it, and then do something about it. Expect others to accept their responsibility as well. If everyone establishes this habit, there will be tremendous power leveraged to identify what went wrong and what needs to be done differently.

On the other side of the ledger sheet are those things that do go well. Again, it is time to establish a habit, the habit of giving credit to anyone but yourself. Remember that things get done not because you say anything. At the end of the process, things get done because people do them. They might be doing them in response to your direction, but this does not detract from the fact that they do them. Give them credit. I was once on a submarine mission when a machine that removes carbon dioxide from the atmosphere inside the submarine became inoperative. The impeller or little fan-like device that moved a chemical solution around the machine had failed; and, because of multiple earlier failures, we did not have another spare impeller on board. With this machine defective, we would have continued our mission, but the carbon dioxide levels would have been high enough to give everyone a continuous headache. Giving headaches is not the way to get people to focus on the many important tasks of the mission. An extremely intelligent and clever crew member, who happened to work for me, did an extensive amount of research and found that with a few minor modifications, he could use as a replacement the impeller of a tiny pump used to remove wastewater from the bowels of the submarine on a routine basis. He modified one of these impellers and installed it in the once-defective carbon dioxide removal equipment. The wayward machine was returned to service and ran like a charm doing its job to its normal high level of satisfaction. Most memorable to me about this occasion was that I recommended to the captain of this submarine that the clever impeller-working gent receive an award to recognize his ingenuity, and the captain refused. He refused because he did not like this gent, for some reason so unrelated and insignificant, that I don't even remember it. How much effort do you think this clever submarine sailor will put into the next engineering crisis that a submarine would have? Look for reasons to give credit. If you can't find any, look harder. Be a real leader and remember that credit is like currency, except that you have an infinite amount of it to give away. Credit can have more positive effects than money.

# KNOW AND NURTURE YOUR PCA

*"If you don't have a competitive advantage, don't compete."*

-Jack Welch

A good leader runs his life like a business. Think about it. Businesses need to make money. Businesses can over-leverage and get in trouble; so can you. You can borrow too much (over-leverage) and get in trouble. Businesses must be acutely aware of their cash flow and adjust to maintain it at the desired rate; you need to be aware of your cash flow and adjust to maintain it at a positive rate; otherwise, no matter how much money you start with, you'll eventually go broke. Businesses need to be efficient to thrive; you need to be efficient to thrive. It takes hard work to make a business succeed; it takes hard work if you are to succeed as an individual. Businesses compete with other similar businesses to encourage customers to come to them rather than to others. You as an individual, have numerous opportunities to compete with peers in your field. The operative words in this last sentence are, "have opportunities" because you don't have to compete. You can come to your workplace every day, do what you think is a reasonable amount of work, and let chance, albeit small, determine whether or not you progress in the business you are in. And as Jack Welsch noted above, you don't need a competitive advantage if you are not going to compete. Again, think of the individual-to-business analogy. For business A to best businesses B, C, and D, business A needs to have a competitive advantage; for you to best the peers with whom you compete, if you choose to compete, you need also to have a competitive advantage. Do you? What is it?

This concept of PCA or Personal Competitive Advantage is a set of terms I developed for myself in the earlier stages of my second career. I reflected on the fact that I had entered the U.S. Navy at the lowest enlisted rank and retired as a commissioned officer with the rank of Lieutenant Commander, responsible for all engineering aspects of a nuclear-powered submarine and all of the people dealing with those aspects. I then entered the civilian world, being hired by a high-standards company working with the commercial nuclear power business, and, as in my naval career, I began as an entry-level employee. Before I left the company, I had achieved the position of Vice President and was an officer of the company, having held several executive and division head positions along the way. The majority of my peers were highly educated, highly experienced, highly motivated employees. So, how did I advance so far and so quickly ahead of them? Having had the opportunity to study the civilian business world at the prestigious Harvard Business School, I quickly concluded that, as discussed above, individuals are like businesses in themselves. I succeeded beyond many others because I had a competitive advantage. Thus, came to be born my concept of PCA, and my conclusion that effective leaders either have or develop a PCA.

Think about it, and if you don't already have a personal competitive advantage, I will share mine with you, and I am confident mine will work for you also if you want it to.

My PCA is a willingness to work. I work harder and longer than anyone else. And I don't do this because I enjoy work. I do this because I enjoy responsibility, and advancing, and

providing for myself and my family, just like any
leader would.

# DEVELOP PEOPLE

*"If your actions inspire others to dream more, learn more, do more, and become more, you are a leader."*

-John Quincy Adams

If you enjoy going to the beaches of Florida for fun in the sun and doing so without carrying a passport and going through immigration control at the airport, you can thank John Quincy Adams at least you could if he had not passed away over 170 years ago. Before serving as the 6th president of the United States, Mr. Adams served as secretary of state, and during his time in this office, he took one of the actions for which he is now most famous. He negotiated the Adams–Onis treaty, an agreement between the United States and Spain. In that treaty, Florida was ceded to the U.S. - and thereafter more readily available to American beachgoers. In addition to being a good treaty maker, Mr. Adams was also a strong proponent of leadership development, and one of his quotes in this realm of interest serves above as an opening to this topic. I call it the Adams test, and it's a simple one you can use on yourself to see if you are the leader you might think you are.

I doubt that any leader would argue that inspiring others is not a key part of her role. However, an experienced leader has a more unique application of this inspiration role, specifically involving the HOW of this task.

Put succinctly, on the day you assume any new position, one of the most important actions you can take is to choose your replacement.

If your choice was a good one, then that person, by your selection, will be inspired to dream more, certainly to learn more, and even more so to do and become more, and you, per John Quincy Adams, will be a leader.

If your choice was not a good one, and we all make mistakes, especially when judging capabilities, then you will also benefit. The chosen one might be unsuited for the job for any number of reasons, but the one most likely is because he/she does not have the drive, the energy, the willingness to work, or the burning desire to take on more responsibility, to develop. Identifying such shortfalls allows you to better understand one of your team. You can then decide if the person can still make a needed contribution in a position of less responsibility, recognizing that she will never rise to the level you first thought she was capable of, or if she should be cut loose from the team so that she can pursue work that is more suited to her motivations and capabilities.

Most importantly in the case of a wrong choice is to quickly move on, choose another for your replacement, and load her up with the kind of work and responsibility that will once again test your judgment.

Why would a leader be so anxious to find a replacement even though one is not yet needed? The obvious answer is to develop that person. Career skills are like human muscles. To grow they must be stressed and stretched. They must even be slightly damaged. The body then repairs or replaces damaged muscle fibers by fusing muscle fibers to form new muscle which is thus increased by growth. With each workout, or challenge in the workplace, the muscles grow, and along with the muscles, the ability to take on more weight and be better suited to take on more challenges. But there are more reasons than this for a leader to shift his work to his subordinates, and some of these might be considered selfish.

A real leader no more gives work to a potential replacement and then sits back and takes it easy than does a trainer tell a trainee to lift weights while he sits back and relaxes with a sandwich and a beer. Real leaders are always self-developing. When the replacement is capable of taking on more responsibility, these leaders are then free to pursue greater challenges that will exercise their own career muscles. Should these more challenging pursuits then lead to an opportunity to move on to a new and higher position, the leader will be well-positioned, having already trained a replacement.

Do we train only our replacement? Of course not. Andy Grove, the legendary founder and one-time CEO of Intel Corporation said it well:

*"There are only two ways in which a manager can impact an employee's output: motivation and training. If you are not training, then you are neglecting half the job."*

WTS would add, *"If you are training as well as motivating, then you are not a manager, you are a leader. "*

The purpose here is not to drag out the same old claims about the importance of training, but rather to emphasize that training, or better said, development, is at the forefront of a leader's mind. There may be a role for academic-style training in a classroom, but the best leaders know that the best training is done by providing experience. I love the story of the boss who was asked if he was going to fire the employee who made a half-million-dollar mistake. The boss replied, "Of course not. I just spent a half-million dollars training him. Why would I fire him and have him take that valuable experience somewhere else?" The point? Train by having people DO things. Team them up with more experienced colleagues and let them first see how something is done and then have them do it themselves. You don't get these kinds of training behaviors in place in an organization by hoping they happen; you get them in place by stating clearly that it is an expectation of yours that they be used - by all employees.

# CONTINUOUSLY IMPROVE

*"Continuous improvement is not about the things
you do well. That's work. Continuous
improvement is about removing the things that
get in the way of your work. The headaches, the
things that slow you down, that's what continuous
improvement is all about.*

-Bruce Hamilton

Nuclear News magazine has pointed out that
U.S. nuclear plants are always improving, and the
magazine substantiated this statement with
several objective measurements that showed
improvements such as power losses decreasing,
nuclear fuel becoming more reliable, worker
exposure to ionizing radiation continually
decreasing, and safety system performance
improving. I can personally vouch for these
improvements. I saw, as well as contributed to,
them during decades of work in the nuclear
power industry. These improvements didn't come
about because of happenstance. They came about
because continuously improving is a way of life
in the nuclear industry, and the work ethic and

general thinking that led to this came, to a large degree, from people who came out of the U.S. nuclear submarine force. Such an attitude is descriptive of a good leader. So, if you want to be such a leader, strive for continuous improvement, in everything you do. My wife and I have some recreational property consisting of about 260 acres of primarily forested land. On the property, we hunt, hike, mountain bike ride, improve wildlife habitat, generally enjoy nature, and teach the young in our family about the wonders of the natural world. Why am I telling you this? To show you, by example, that the concept of continuously improving can be used in just about any endeavor. We use the principle in managing this property. We have established metrics to measure improvements, such as the acreage of wildlife feeding areas, the number of food-producing trees we have planted, and the number of wildlife sightings. We benchmark by visiting other properties. We have long-term plans and strategies well laid out on paper and use these to develop and track an extensive number of work items on an annual action item list we develop each year. Actions completed and results achieved are recorded, analyzed, and used to develop future actions.

But again, some cautions on the application of continuous improvement efforts. Do not rely on continuous improvement advice found widely on the internet. There is more such advice available than politicians have excuses for not getting things done, but much of that advice is wrong. Internet articles carrying this advice often employ non-results-oriented terms, like "model," "tool," "process", and "program". The articles are often linked with other advice-laden, vendor-pushed, do-this-and-get-better sales pitches, with a variety of names, usually all having some flavor of "Quality Improvement." No doubt some have found these programs to be useful, depending, of course, on how well they are implemented. But a weak point of these programs is just that they are "programs." Putting a program in place can give one the euphoric feeling of having accomplished something, when, in fact, nothing might really have been done to fix a problem, or otherwise improve performance. Patient safety in our healthcare institutions is a good example. The Institute of Medicine has reported that 98,000 patients die from preventable medical errors in U.S. hospitals each year. Most interesting about this statement is that it was made twenty years ago. Today the number of patients experiencing harm in hospitals ranges from 250,000 to 400,000 each year. And this human tragedy is not limited to patients. It includes the healthcare staff as well. OSHA reports that "more workers are injured in healthcare and social assistance industry than any other." In 2019, U.S. hospitals reported a rate of work-related injuries and illnesses almost twice the rate of private industry as a whole, leading OSHA to report, "A hospital is one of the most hazardous places to work." Think about the safety trends reflected here. This performance is after more than fifteen years of safety improvement programs. What's the point here? The point is if you want to be an effective leader, meaning you want to continually improve your performance, FOCUS ON RESULTS-ORIENTED ACTIONS. Such actions are the engines of change.

It's also important to make time to think about what improvements are needed. I once had the opportunity to meet an elderly gentleman who was the owner of a large custom-made furniture manufacturing business. He loved his work and proudly showed me around his manufacturing facility. He was 84 years old, and still worked 11 hours a day, six days a week. He understood his business right down to the ground level. He had a special sewing machine for those times when he went onto the upholstery floor and worked the machine himself. It was designed to run more slowly than the usual sewing machine, which he said was necessary because of his age. I relate this fact only to help the reader understand what kind of a person this gentleman was, a hard worker, a get-in-the-field, Mustang kind of guy, and one with a lot of practical sense gained through years of hard work. As we walked, he dropped pearls of wisdom like a tree does leaves on a late fall day. One of those pearls was his view of being an independent businessman. He told me he loved owning his own business, because, "I am my own limitation". He was right, I quickly concluded. If something was too hard, he had no one to complain to. If it was too hard, it was his fault. He was the one who should make it easier to come up with better approaches to the problem or whatever it was that seemed "too hard". As mentioned earlier in this book, a good leader runs her life like a business, and as such, she is her own limitation. Are your work hours too long so that you're shortchanging your family? Do you find yourself doing work that others should do but don't because they don't have the motivation or the skill that you do? Do you find that you are so bogged down with paperwork that you can't get out in the field and see what the employees are doing? If any of these or anything similar is true, you should complain. Where do you go to complain? A mirror. Any mirror. These are problems of your making. So, stop complaining and get them fixed. As Bruce Hamilton, the award-winning TV, radio, and newspaper journalist, said in the opening quote of this chapter, continuous improvement is about removing the things that get in the way of your work.

The above may seem snide, but it makes the point that you are your own limitation. You can wait until problems like these occur and then react after you recognize that you are the one who is supposed to fix them, but better yet, you can take action to not let them happen in the first place.

Try this exercise. For starters, do it once a week. You might find that is too frequent, so let experience be your guide in making adjustments. Set aside some time, maybe an hour. Make sure you are in a location where you will not be disturbed. Then, think about what is going on within your work environment that you don't like. And then think about what you can do to change it. I have done this exercise numerous times. In one case, I thought I had too much work to do and that did not allow me enough time to think strategically. As a result, I recognized that even though I was delegating, I was not stretching my employees enough. Some of the projects I had thought might be too difficult for them I put on my own plate. I was wrong, completely wrong. Many of my employees were more than capable of doing some of these more challenging assignments. And by not letting them, I was shortchanging their development opportunities as well as overloading myself. In another instance I found that I was overloaded with emails from my subordinates, many, honestly, just to keep me informed, while some were just a sneaky way of telling the boss what they were going to do and then assuming if I didn't object, that was permission to proceed. My "me time" thinking led me to a great fix. I put out the word that I was no longer reading my emails, and nothing could be assumed by the fact someone had copied me on an email. I further said that if something was important enough to inform the boss, it was important enough to inform him face to face. My email problem went away.

# LEAD THOSE BELOW, BESIDE, AND ABOVE YOU

*"The function of leadership is to produce more leaders, not more followers."*

-Ralph Nader

Any effective leader is humble. This is particularly important in leading those beside or at the same level as we are, and also those above us in the chain of command, to use a few military terms. However, the trait of humility is also important in leading those below us in the organization, and even the term "below" needs to be used with caution. Those we are fortunate enough to lead are not below us in any way other than perhaps in their position on the organizational chart. This is a mindset that those in charge sometimes let escape from the corral of their minds, and this mistake can be performance-wise fatal if they have any interest in encouraging the teamwork that is essential to an effective organization. The people you lead must see you as a person, as one of them, as one who can relate to their problems and challenges. If you are a

Mustang, this will be easy, but there are many ways to achieve this. Start by showing humility. Talk to people, not just about work, but about life in general. Let them know that you face the same kinds of problems they do and that you have the same exposures that many of them do. Most importantly in the humility arena, when you make a mistake, admit it. Again, you are human, just like them, and although as a leader it is our job to minimize mistakes, just about everyone makes them. Own it, learn from it. Behave the way you would want them to behave when they make a mistake. Again, remember Exampleship.

Regarding leading those above us, several texts have been penned on the importance of "managing up" or "managing your boss". Unfortunately, these terms, although often not intended to, can give the impression of subterfuge, or imply the use of devious methods to inappropriately influence one's boss. I recognized the inappropriateness of these terms some time ago when I realized I wasn't willing to share them in any conversation with my boss. If the concept is a good one, and it is, then why not be transparent about it? Before answering this question, determine first whether or not you are indeed being devious or doing something inappropriate. Ask yourself the question, why am I trying to manage up? If the answer is, to benefit yourself, be it in recognition, advancement, or other compensation, then you are indeed trying to inappropriately manage your boss. But if the answer encompasses the good of the company, the project, or the mission, then influencing the boss is something for which he should be grateful.

How do you lead your boss? There are three essential elements in the answer to this question:

1.  Never forget that the boss is the boss. The best way to do this is, without fail, to show respect.
2.  Never miss an opportunity to make the boss look good.
3.  When you want the boss to move in the direction you want him or her to move, show the effectiveness of the actions you desire the boss to take. If you can't show the effectiveness, maybe your idea wasn't so good after all. If you do show effectiveness, provided you've done numbers 1 and 2 well, you will now likely be leading your boss. But remember, even if you have done all this

exceptionally well and the boss will not be
led, the boss is always the boss.

# FOLLOW-UP TO ENSURE EFFECTIVENESS.

*Diligent follow-up and follow-through will set you apart from the crowd and communicate excellence.*

-John C. Maxwell

John Calvin Maxwell, the American pastor and author of numerous books, primarily on leadership, unfortunately captures the thought well, in proclaiming that if you are good at follow-up, it will set you apart from the crowd. I say, unfortunately, because the "crowd" or, said differently, so many, unfortunately, do not follow up. Believe me. If you are an effective leader, you will not be among that "crowd."

In the world of improvement, which is one in which all should operate, a good assumption to make is that everyone has problems. If you think you don't have problems, then that in itself is a problem. As Admiral Hyman G. Rickover, the founder of the U.S. Navy's nuclear fleet and not one to beat around the proverbial bush with his words, once said, in dealing with problems, optimism is synonymous with stupidity. Also, although not as prestigious a name as that of the Admiral, but with a lot, and I mean a big lot, of management miles having passed under my feet, as I discuss in *The Observant Eye*, these problems often continue, not because management doesn't know how to fix them but rather because management often doesn't know that the problems exist. I made my living for many years identifying such problems for organizations and then describing them to management along with providing sufficient helpful insight into the causes of the problems that allowed the managers to go after and address them, which they generally did. Unfortunately, sometimes, after I had left some organizations, the problems had come back. The reason for their return was that although managers took action, they often didn't FOLLOW UP to ensure the action they took was effective and long-lasting, or even more so, they made sure there was some effect of their actions when they were applied, but made little or no effort to relook at the effectiveness after some time had passed. Many problems and actions taken to address them cause a lot of "buzz' in an organization. The workforce sees management action and this is like street news. Many pay attention to it just because it's something a little different. It carries a good bit of curiosity along with it. But all of this organizational excitement goes away in a short time. People often even forget what happened and what action was taken, and it is at this point that the effectiveness of the corrective action taken must stand on its own and sometimes doesn't. Good leaders have a simple model related to this imprinted in their minds. It looks like this:

PROBLEM ID & CORRECT CAUSE; APPLY FIX; CHECK PROBLEM GONE; LATER, CHECK FIX STILL IN PLACE AND PROBLEM STILL GONE

# MAINTAIN THE VALUE OF YOUR WORDS

*"Good words are worth much and cost little."*

-George Herbert

The quote above, from Mr. Herbert, who was an English poet from the 1600s and known for the purity and effectiveness of his words, is a statement itself on his effectiveness with words. It makes the point of this entire chapter and does it with eight words.

We can probably all think of a colleague whom we do not particularly look forward to listening to in work-related discussions, and there is a wide variety of reasons for this. The person might be overly wordy, tending to ramble, or might often convey information that leaves us wondering what their point is. However, in general, if we care to not listen to someone, it is usually because we do not value their words. Remember this; think about how effective a leader can be if people don't value their words; choose your words wisely; and whenever you speak, set a personal goal for yourself to have people listen. Mustangs will generally be strong in this area for the simple reason that they have been on the receiving end of unclear messages. But one need not have once been at the working level, or a Mustang, to develop this strength.

There are several things to do in striving to achieve the goal of having people listen:

- Be stingy with your words. Keep the thought in mind that those in your audience are as busy as you are (even if you think they are not) and have little time to waste. Don't open your mouth unless you have something valuable to say. Ask yourself if you would appreciate hearing what you are about to say.
- Organize your thoughts beforehand, if at all possible. This is back to the earlier section on organizing your thoughts to communicate. Do this by writing your thoughts down. Be critical of them. Ask yourself what it is you want the listener to take away from your words. Then relook at your words and ask yourself if there is a more effective way to communicate the thought. This may sound like a lot of extra work. It does to a lot of people and that's why they don't do it. But they should.

- One of the most valuable tools one can apply in choosing words is a question asked to yourself. So what? If additional words have to be used to answer the question, your first chosen words were poorly chosen. Never make it necessary for someone to have to ask, So what? Ask yourself this question first and then add your response to what you had originally intended to say. For example, if you tell someone that the healthcare community does not have anything like the airline industry's National Transportation Safety Board, you are communicating an important piece of information. But if you want to make that information even more valuable, answer your own "So what?" question in advance, and add on something like the following: As many as 400,000 people die each year from preventable medical errors and these events often go without critical examination.
- And finally, provide insight. People will appreciate it. Insight adds value to your words. What is insight? Webster tells us that insight is the capacity to gain a deep and intuitive understanding of something. Isn't that what we want, i.e., to have our listeners gain a good understanding of what we are

saying? One good way to think about insight and include it when appropriate is to recall what Phillip Kotler, the American marketing author, consultant, and professor emeritus, said, "Great insight comes from seeing something odd and finding out why." Find out why, and include it in your remarks. Malcolm Forbes said the best vision is insight.

# BENCHMARK PROACTIVELY

> *"What a business needs most for its decisions especially its strategic ones is data about what goes on outside it. Only outside a business are there results, opportunities, and threats."*

-Peter Drucker

In attempting to either preach about a problem in your facility or your organization or even to just have them perform better, an excellent first step is to go and look at how some organizations that do not have this kind of problem or are performing better than yours do business. This, of course, is called benchmarking. Many organizations do benchmarking. Unfortunately, not many organizations do proactive benchmarking. This means you visit other organizations and learn from them before you have problems, and "before" is the most important word here. You are just trying to get better, maybe even move from the good to the excellent category. Although it certainly wasn't called benchmarking in the nuclear submarine force, we essentially did the same thing with a different approach. One of the most active initiatives on the nuke boats was assessment, both self-assessment and assessment by those who are essentially one's peers. This means that someone who does or has done what you are currently doing comes into your organization and offers their perspectives on your operations. Because you are essentially benefitting from the experience of others, this activity has the essential elements of benchmarking. Also, self-assessment is done frequently and rigorously in the nuclear submarine world. To not do self-assessment when living in a world where outside agents are regularly coming in to assess the performance of your organization would be foolish. And what is a key element of self-assessment? Getting some involved who are from other organizations so that they might uncover blind spots that your workforce and management might have.

The belief in and strong support of assessment, by both self and by others, is another one of those traits that have been carried into the civilian nuclear power arena by those who have grown up with it in the Navy Nuclear program and then, following separation from the military, went on to operate nuclear plants that didn't drive submarines but that generate the electrical power that lets us read at night and take hot showers.

One other element of the effectiveness of looking at others to find out how to do what you do and do it even better, whether you put it in the benchmarking or the assessment categories, has its roots in the term Mustang. Get the troops involved. Certainly, a manager or supervisor can learn much from seeing how others do business, but those who operate the valves or drive the machines or interact with the customers, i.e., the troops, can also learn, not only for themselves but, as well, for the good of the company. I have taken working-level personnel on numerous benchmarking trips and it was a pleasure and a learning experience for me to see their reactions. There was a strong sense of self-determination on their part, a determination that they could do at least as well as and maybe even better than those they observed. And they tended to see things that a manager or supervisor might not see. I cannot advocate more strongly for including some of those in the workforce on these, let's call them learning, trips.

Again, looking at an operation and comparing it with how that operation is done in other organizations might be called assessment or benchmarking, but whatever it is called, in effectively run operations, it is done proactively, to improve, not just when a problem occurs. Thus, as indicated in the opening of this section, an effective leader benchmarks proactively.

# BE WARY OF INITIATIVES

*"If you spend too much time learning the 'tricks'
of the trade, you may not learn the trade. There
are no shortcuts. If you're working on finding a
shortcut, the easy way, you're not working hard
enough on the fundamentals. You may get away
with it for a spell, but there is no substitute for the
basics. And the first basic is good, old-fashioned
hard work."*

-John Wooden

Initiatives can be good, but they can also be
distracting. Unfortunately, many managers love
initiatives. Launching initiatives gives one a sense
of doing something, taking action, and making
progress. However, over time, there is a natural
tendency to lose focus on the basics as new
initiatives are added. New initiatives garner the
attention of both the workforce and the leaders
but in doing so, they subtly displace attention to
the basics and move them to a back seat. For
example, the basics of running any complex
equipment include having good procedures;
having components properly labeled; following

the applicable procedures; monitoring the operating components; communicating clearly; and identifying and correcting problems. If these basics are followed religiously, the operation will do well. A special effort to spiff up the surrounding work areas might be initiated, and that by itself would not be a problem. It could even be a real strength unless, for whatever reason, it causes the equipment operators to lose focus on the basics. This is when initiatives become a problem.

A periodic "back to basics" effort is worth the time and effort. A good leader never loses focus on the basics. Even when initiatives are desired or needed, leaders never take their eye off the basics.

# HOLD PEOPLE ACCOUNTABLE

*"At the root of poor performance in most organizations that have such performance is a lack of accountability."*

-WTS

In January 2019, The Predictive Index™ surveyed 156 CEOs, presidents, chairmen, and chairwomen. They asked a slew of questions that cut to the heart of what drives these people, what their challenges are, and what keeps them up at night. Their answers revealed the patterns of high-performing CEOs and allowed an exploration of the executives' inner thoughts and biggest weaknesses. A key finding that came out of this study is that the executives' biggest admitted weakness was not holding their people accountable.

Based on my experience in conducting hundreds of management assessments of large companies, I was not surprised at the above finding. Lack of accountability is generally the problem when an organization has just about any performance problem. Admiral H.G. Rickover said it well when he said: If responsibility is rightfully yours, no evasion, or ignorance or passing the blame can shift the burden to someone else. Unless you can point your finger at the man who is responsible when something goes wrong, then you have never had anyone really responsible.

I was surprised that the above executives self-identified their problem, and I credit them for doing so. You can't fix a problem you have if you don't know or won't accept that you have the problem.

There is a wide variety of reasons for the lack of accountability being such a prevalent issue in many organizations. None of these are valid. Let's talk about a few of those reasons. One is that people often don't have a good understanding of what "accountability" is. This shortfall is not helped by the large number of "management' books that contain statements like:

"Accountability is when a person accepts responsibility for their actions and decisions"

and

"Collaboration encourages greater accountability".

Both of these statements come in with a grade lower than that of inaccurate. That, of course, would put them into the category of BS. They are more troublesome than just being BS. They hinder the implementation of accountability by contributing to the misunderstanding of it. Having coached many supervisors and managers on how to hold people accountable, I found the following simple, self-developed definition to be most useful:

Accountability means that there are consequences for actions and results.

If you, as a company representative (at least in my company) use foul language in front of a customer and thereby present an unsavory picture of the company, that is an unacceptable action, for which there will be consequences. You will have known this was an unacceptable action because I will have told you my expectations when you first came into my organization. The consequences might be limited to me giving you direct feedback on the occurrence, appropriate if this is a first such offense. Or it might be moving you out of your position. The consequences might differ, depending on the details of the occurrence, but there will be consequences. Such is called accountability consequences for actions. Results are another product that raises the issue of accountability. People, except for those in the government some would argue, are paid for results. Regardless of the actions you are taking, if you do not achieve results, you are and should be held accountable.

If I tell you, as one of my direct reports, that I expect you to complete a job by noon tomorrow, and then tomorrow, when the hands of the clock are both held high in the air, you have not completed the job, there will be consequences for lack of results. Again, the degree of consequence is to be determined by the details of the shortfall, but there will be consequences. There will be accountability.

This is a simple concept. The only tools needed are clearly stated expectations beforehand, and candid feedback afterward.

Another reason for the lack of accountability is over-empathy. You can relate to where the employee is. You have been there yourself. You know the job is difficult, in fact, possibly too hard. Such is the thinking that leads you to conclude you should lower your expectations to meet whatever performance the employee has offered. The downside of this, of course, is the job doesn't get done to the standards you had expected. Additionally, it gives short shrift to the abilities of the employee. It assumes the employee is not capable of meeting the expectations set. No one benefits from this approach.

Another invalid but popular reason, sometimes not even being recognized for what it is, is a lack of professional distance. The employee is your friend. You have known him for years. Your families may go to church together or perhaps you and the employee often golf together on the weekend. It is also possible, even likely, that not that long ago, you and the employee worked together at the same level in the company hierarchy. Because of the special traits you have demonstrated and the additional effort you have put into everything you do, you were advanced in position. Now your friend works for you. Seems like a problem, doesn't it? Well, it is not. Again, give the employee some credit. She is a professional, just like you. She understands professional distance. She can separate a social relationship from a professional relationship. Keep in mind that this employee is a team member, not a family member. She expects to be held accountable for what she does or does not do. If she doesn't, she shouldn't be working in your company anyway.

And finally, maybe you feel that you just don't have it in you to hold people accountable. Your nature is not to be the bad guy. You want life to be fun, and everyone to be happy. I have seen an uncountable number of examples in which a person in charge was reluctant to directly tell an employee he was not meeting expectations and, as a result, an opportunity to change performance for the positive was missed. For example, I once watched an employee being observed by a supervisor to ensure he was properly performing "peer checks" of another employee during an evolution that required sequentially operating a number of switches. The peer check required the performer to read aloud the nomenclature of a switch that was specified in written guidance, point to the switch, and read, again aloud, the nomenclature on the switch itself to ensure it was the same as specified in the written directions. The peer checker was to acknowledge aloud both that the nomenclature specified in the procedure was read as stated and that the nomenclature on the switch matched that in the written directions. Following this, the peer checker was to verify the operator's hand was on and about to operate the correct switch. What actually happened was that the peer checker acknowledged what the performer was doing at each step of the evolution with only a slight, barely perceptible, nod of the head. The supervisor, rather than clearly state what expectation the peer checker was not meeting, instead said only, "It would be better to verbally acknowledge your agreement with each step." Although an accurate statement, this dilution of the point that the expectation, which was clearly defined in the administrative procedures, was not met. This shortfall in the application of accountability often stems from the misconception that accountability has to be more negative than objective. It does not. The principle is simple: There should be consequences for one's actions. In our example, the actions were not in accordance with procedures. The consequences needed only to be that the checker was confronted with the fact that he did not do as expected. If you accept this but still conclude that you cannot deliver or ensure delivery of the consequences, as in our example, you likely are not one to be in charge.

If one only applies accountability in a negative sense, it is often said, in those "management" books containing other less-than-useful or even misleading information, then fear and anxiety will permeate the work environment don't believe it! One can apply the simple definition of accountability that I provided above to positive interactions as well. If an employee performs some action or achieves some result that you, as the boss, think is highly positive, you can and should provide consequences in terms of reward or recognition. So, the definition fits! But I strongly encourage not using the accountability term in this way. I suggest, alternately, to consider this later case to be one of recognition, appreciation, and positive feedback. Leave the "A" word out of it. I say this because of how widespread accountability shortfall problems are in our country. In the 1980s, for reasons I do not know, there was a shift in our business communities to managing popularity. This was the beginning of the decline of real accountability and it seems that many have never recovered from this slide. So, if you start using the term in dealing with positive situations or interactions, it further adds to the potential confusion and weakening of the real concept of accountability, which is geared toward improving performance and making things better.

One other aspect of accountability is rarely talked about, self-accountability. Bear with me on some potentially boring technical details to better understand an important point on self-accountability. A submarine's pressure hull is shaped like a Good & Plenty candy or a medication capsule. This hull is surrounded by large tanks, called "ballast tanks" that have openings in the bottom and valves on the top, called "vents" that are normally closed. To submerge, one simply opens the vents, the ballast tanks flood, and hopefully the sub submerges, but not too far. If the boat has negative buoyancy, it could sink to the bottom with catastrophic results. If it has positive buoyancy, the results are worse, because worse than death would be the embarrassment of not being able to submerge the boat. Death can, in such circumstances, be a real possibility. Think about it. An enemy surface ship is about to put a warhead into the side of your vessel; you have, you think, the option of escaping by submerging your vessel; you try and fail. The boat will not submerge. Bang! You're dead. The ideal in submerging at the proper time and in the proper way is a result of perfect or near-to-perfect neutral buoyancy, in which case the boat submerges and then like a plane pushed along by its engine, the bow and stern planes act like wings to harness the upward or downward force of the water and drive the boat up or down through the various depths of the sea. At one point in my submarine career, it was my job to determine how much water to place in what tanks to achieve this neutral buoyancy when the ballast tanks were flooded. I did this by having calculated the weight of all of the crew members, and adding to this the weight of all the spare parts, food, weapons, and other supplies we had taken on board, then combining this weight with that of whatever amount of water would be taken on in the flooded ballast tanks when the sub dived, and counterbalancing it with the unflooded volumes of air in various other tanks and compartments. These were quite complex calculations, with numerous opportunities for error. Can you guess how many people checked my calculations? If you guessed any number other than zero, you would be wrong. In my earliest times in this job, I was surprised that no one was going to check my work. Extra checking was not a luxury we had the good fortune to have. We typically did not have a lot of people sitting around, waiting to redo the work someone else had just done. It was a lesson that drove home to me the concept of self-accountability. I am confident that had I known that others would be checking my calculations, I would not have paid the level of attention to detail to them that I did, recognizing the life-and-death importance of being right. Employing this self-accountability and enjoying the pride that came with it, I got to be quite good at these calculations and often, upon submerging, would go to the Control Room and watch in silent pride as the boat went down and then just hung there like a picture on a nail, in a state of perfect neutral buoyancy, a testament to the accuracy of my calculations.

Accountability is the oxygen in an organization's atmosphere, but if you want to be confused about accountability and its use in practical terms look in a dictionary. Look for the definition of accountability anywhere, including within the uncountable number of articles discussing it on the web. What you find will mislead you. The biggest shortfall I have found in those discussions and definitions is confusion between the discussion of accountability and that of responsibility. Let me provide an example that is based on having seen this problem in numerous organizations. Let's say I have a new employee, John, and I tell John in very clear terms, "John, you are responsible for keeping this stairwell clean (or any other task you wish to apply to this example). Several days later, I observed the stairwell was filthy and covered with paper trash. Suppose I neither say anything nor do anything to John. Is John still responsible for the staircase's cleanliness? Of course, he is. I made that quite clear. Have I held John accountable for the unsatisfactory execution of his responsibility to keep the staircase clean? I have not. So, I would argue that John is responsible for the task but not accountable. Rather than get into a debate on word usage, let me share my experiences with a multitude of organizations, which has led me to conclude that at the root of poor performance in most organizations is a lack of accountability. And I say this because, in the world of a real leader, accountability is consequences for actions. These need not be drastic consequences. I need not fire John because the stairs are dirty, but I do need to react in such a way that John recognizes there are consequences for him not adequately executing his responsibilities. I can do that just by talking to John, as long as I make it clear that he did not meet my expectations and, therefore his performance is not satisfactory.

It is human nature to want to be nice, and it is because of this tendency that holding people accountable is one of the most frequently occurring shortfalls in underperforming organizations. To deal with this apparent conflict in human nature, one only needs to recognize that accountability, properly used, can be good for the person being held accountable. It can help them by motivating them to do better. I have seen many organizations in which a reluctance to use accountability becomes the normal way of doing business. One supposed leader sees another not hold people accountable and thinks that is the way things are done, thus following suit. The shortfall then becomes a way of life in that organization, and a major cultural change would be needed to move toward greater accountability.

# DON'T STRESS ABOUT WHAT YOU CAN'T CONTROL

*"Recognize those things over which you have no control and don't waste time and energy worrying about them. Save your supply of worry and act upon those things you do have control over."*

-WTS

The approach of not stressing about things you can't control, if practiced enough, can become a habit, a healthy and productive habit. Think about it. Things get done because people do them or cause them to happen. Behavior is essentially defined by a person's actions. If you want something to happen, someone must act to make it happen. If things are not getting done, people's behaviors must change, or those things will never get done. And remember, getting things done is a primary function of a leader. The premise of this principle is that there are things that you can't control. If indeed you can't control them, then of what value is the action you take to try to control them? This is a simple principle, maybe overly simple, and that might be why some people in charge don't adhere to it. They don't think about it and overlook the obvious. There is no action in getting stressed, and only action can get things done. If the stress causes you to do something, some would argue that this is a good thing, but they again don't consider the premise. You have no control over the thing about which you are getting stressed. And there are downsides to getting unnecessarily stressed. First, it might not be good for your health, especially If getting stressed becomes a chronic happening. It also is not good for your happiness. I am a strong believer that happiness is one of the most important factors in one's professional life. Happiness is linked to job satisfaction, to a willingness to take on challenges without fear, and to come to work each day with the best attitude you can. But more directly to the point of Nuclear Mustang Leadership, the second reason for not getting stressed about the things you can't control is that stress takes energy, and a real leader can't afford to use, or should I say waste, that energy. There are too many more important things to do, and actions to take, to address those many things over which you do have control. I have coached a number of my direct reports on this principle, and these are people for whom I have the greatest respect, and from whom I learned more than time and space here allow me to describe. It surprised me that they had not come to this same conclusion on their own, but they hadn't, and when I suggested it, they quickly got it and thanked me in various ways.

As the Chief Engineer on a nuclear-powered submarine, I once walked into the central control area for the reactor while the submarine was in port. Even in port, the reactor must be cared for, and that care requires having a large source of electrical power. This power is normally brought onto the submarine via large cables coming from the shore. In emergencies, when that power from shore is lost, there is one more option, that being an emergency diesel generator, but this is your last option and, to simplify the discussion, bad things can happen to the nuclear core of the reactor if you lose this last option. As I walked into the area, a senior reactor technician in charge at the time was in an obvious state of stress, almost frantically looking at gages and other instruments, and lamenting to no one in particular

that we were in trouble, that we had lost the
source of electrical power from shore, a situation
over which he had no control. The emergency
diesel generator had started, to provide that last
available source of power to the equipment that
cooled the core. He was worried, extremely
worried, that the diesel engine would stop, for
whatever reason, electrical power would be lost,
and the reactor core would melt, a terrible event
begun by something over which he had no
control. He was worried about an eventuality that
might or might not happen, regardless of
whatever he did. At the same time, he was so
absorbed with his worry that he wasn't
considering all of the many things related to this
predicament that he did have control over, and he
was doing nothing to act on those items. He was
essentially frozen in fear, worried about
something over which he had no control. As I
entered the area, he blurted out to me, "Do you
know we're on the diesel?" Rather than criticize
him for his lack of action, I chose instead to break
the tension with a bit of humor; I looked at my
feet and asked, *We are*? It was not that I was
taking the loss of shore power lightly. I purposely
used humor as a means to release this gent from a
mental state that was limiting his thinking. Of

course, I thought losing shore power was not a good thing, but my extensive experience and training had opened the floodgates of thought now flowing into my mind about the things that would ameliorate the situation and that we did have control over and that should be done. This humor might not have drawn laughs from a barroom crowd, but it did get this gent to snap out of his state of worry so he could draw on his extensive knowledge, and recall the many things over which he did have control, such as ensuring that sufficient fuel and other support materials were made available to help the diesel generator should it be needed for a longer term than anticipated. My unstressed approach helped to lighten the mood in this situation, thereby helping him to relax, at least a little, enough to think more clearly. He immediately reverted to a highly professional demeanor and began spouting directions to several other people for actions to take. From that point on he did an excellent job, as he had so many times in the past, under different circumstances. The bottom line is: don't stress about things over which you have no control. It won't help, and it could well degrade your ability to act on those things over which you do have control.

# BE RELIABLE

*"A man who lacks reliability is utterly useless."*

-Confucius

Be reliable, meaning that if you say you're going to do something, do it. This seems like a commonsense statement, and it is, but then why do so many people not follow this advice? I am particularly sensitive to the importance of reliability because of my days in the enlisted ranks of the U.S. Navy submarine force. My bosses were the officers. If something needed to be done at any of the many levels above the workforce, it was usually an officer's actions or directions that made that action happen. And those officers, at least most of them, were good, dedicated people with the best of intentions. Like most people, naval officers want to do what they say they are going to do; they just sometimes don't do it because they get so busy with other things to do. Some of these same people also are generally not very good at prioritizing what they have to do. Prioritizing, like doing what you say you are going to do, is also not hard to do. I have found the solution to the issues of reliability and prioritization to be the same, and that is, to use lists. If you say you are going to do something, essentially you are committing to whomever you are speaking. Commitments are important, so put them into a tracking method to be sure they get done. Even write them down; put them on your smartphone or computer, and just make a list, unless you have some other means of tracking. Having a list of what you have to do allows you to track what has to be done, but it also facilitates prioritizing. A quick look at a list that shows all of your more important planned actions lets you see an item you committed to doing relative, in importance, to other things you have to do. If reliability is important to you, then it is clear which item gets done first. I often keep a 3 X 5 card in my pocket with my planned actions for the day. If I say I'm going to do something, it goes on my list. If I intend to do it at a later time, it goes off my daily list but into another tracking system to ensure it gets on my daily list at a later time. I have been recognized for my reliability on many occasions in the personal feedback I have received over the years. This kind of feedback is worth analyzing, but that's for a later discussion.

# HAVE A PLAN

*"Plan your work. Work on your plan. If you don't have a plan, plan to fail."*

-WTS

The value of planning is often overlooked in discussions of leadership. Why? I don't know. But I do know that I have been successful in many endeavors, and in just about everything in which I engage, I have a plan. An effective leader is recognized for getting things done, and if you want to get things done, have a plan.

Nuclear submarines are run hard when operating. They are in a harsh chemical environment being surrounded by highly corrosive salt in sometimes exceptionally cold water, exerted on by remarkable forces caused by extraordinary water depth, and periodically exercised in extreme ways as the submarine carries out its appointed mission. When the submarine comes into port, there is typically an extensive amount of work to be done to return the craft to its optimum operating condition. Essentially all of that work, at least all of the

important work, is "planned". I have been a
Planner for submarine work, and I have been in
charge of a division of submarine planners.
Planners are the people who develop the "plan" to
get things, like work, done. They provide written
instructions; refer to specific maintenance
manuals and other technical guidance when
needed; and generally, write the instructions for
doing whatever work is to be done. I know the
value of a good plan. A well-constructed plan
improves the efficiency of the organizations
implementing or supporting the plan. Such a plan
coordinates the various organizations and
resources such that the right people show up at
the right time to do their assigned work. The plan
defines not only what is to be done, but who is to
do it; when it is to be done; how it is to be done;
where it is to be done; and even why it is to be
done, all to provide the maximum insight to those
who will be implementing the plan. All this is not
to say that a would-be leader should first become
a nuclear submarine planner, but rather to
emphasize the importance of planning. An
effective leader will have a plan for every day of
his or her life. As I sit here writing this, my daily
plan is lying to my left. Using a plan is a habit
that develops. My current plan is an outline of

everything I need to get done today. It has fourteen items on it. Priorities are assigned such that the most important items are designated. Just a glimpse of the list shows me which items need to be done before other items are to be attacked.

Are you about to perform a very special evolution? Maybe it's special because it has the potential to hurt someone. Maybe it involves a lot of money or very expensive equipment that could be damaged. Maybe it's special just because it's important to your project, your company, or your family. Whatever the reason, if the evolution is special, it likely warrants extra steps in advance, i.e., planning.

My submarine mates (I'm sure) and I would encourage a few key elements to be addressed by your plan:

First, before starting the task or evolution, ask what is the worst thing that could happen.

Second, put an element in your plan to prevent the above from happening.

Third, make sure your plan has actions to take if it does happen.

Fourth, put yourself in the position of overseer, and think about what questions you would ask if an event happened during the task. What kind of questions might be asked of you if it happened? Ask those questions now, before the activity starts, and include the actions resulting from the response in your plan.

Fifth, make sure your plan has well-defined expectations for RESULTS. Why are you even doing this plan?

Sixth, when you think your plan is ready, sit back and take a critical look at it to see if you have an obvious focus on the performance of people. One can fix or write all the procedures one wants, and build the best and most well-supported facility in the world, but facilities and procedures do not get things done. People get things done. Focus on actions that will positively affect the performance of people.

Seventh, the plan should employ proven, and I emphasize proven, concepts, such as self-assessments, and walk-throughs of procedures before they are performed.

When you execute your plan, achieve the highest level of success or learn what could have been done better. Then catalog that and any other lesson learned, and file the learnings away for the next time you perform this or a similar evolution.

# ENSURE PROBLEMS GET FIXED EFFECTIVELY

*"It might be done, but is it done-done?"*

-WTS

There is not a great deal of literature on corrective action, yet a good leader recognizes it is one of the most important elements of any organization wishing to continually improve its performance. What is corrective action?

Suppose you're walking back to your office, carrying a cup of coffee, and you drop the cup. The broken cup and spilled coffee on the floor, in the context of the discussion in which we are about to engage, can be called a deficiency, or more simply, a problem. Problems require corrective action, and in this simple example, the corrective action consists of you, or someone, getting a few rags or paper towels and wiping up the mess. Suppose you see a problem, but your schedule and priorities do not allow you to take corrective action. An example might be the lack of cleanliness at the coffee station that you noticed as you got your coffee. The unsanitary condition might have some health implications, but it also paints a less-than-desirable picture of your company for any visitors. The problem can't be fixed right now, but perhaps you could write yourself a reminder note to either come back later and address it or ask someone else to address it in the near term. These examples are of conditions, but problems or deficiencies need not be physical conditions. A deficiency could be a performance issue or a behavior. If you think about it, you will realize that we all have deficiencies. People have them, places have them, organizations have them. And if we are continuing to try to get better, we are continuing to take corrective actions. Corrective actions are like steps on the ladder to success. Keeping a record of deficiencies not only ensures they get addressed but also allows one to analyze the deficiencies and look for common causes.

One other regularly found aspect of corrective action and the problems to be addressed is that the problems get fixed but not for the long term. I have found this to be the case so frequently that in the organizations I led, I began the concept of "done-done", a concept that would serve any would-be leader well. Typically, in my organization, a problem would occur; our conscientious management team would fix it and put a process in place or take some action to make sure it doesn't happen again and report these actions as "done". The reporting person would soon be challenged by his peers, "done-done"? The latter means not only was the process put in place but follow-up was also done to ensure the process has been effectively implemented and achieving the intended results. This grouping of words, "effectively implemented and achieving the intended results" is just a different way of expressing "done–done", a term and concept of exceptional value to a leader.

A side value of having a record of deficiencies and corrective actions is that such a record allows periodic matching of corrective actions to problems those actions are intended to fix. I have reviewed hundreds of corrective action records and programs, maybe even thousands, and I saw no more frequently recurring shortfalls in these efforts than a major mismatch between deficiencies and the actions taken to correct them. Often corrective actions include putting in place programs, processes, or procedures. The intent of this effort is sound. The thinking goes like this: if I just fix the problem, it might happen again, but if I develop a program or process or procedure that will take care of the problem this time and any potential for it to happen in the future also. Unfortunately, this thinking does not go far enough. It is not programs that correct problems; it is the IMPLEMENTATION of those programs. Herein lies the term found in the uncountable number of issues I have identified in this area implementation.

It is doubtful that any individual is going to have the time or desire to set up a full-blown corrective action program with detailed records. But the point is for an individual to remember the key element of this topic on which to focus. In this discussion, that element is implementation. When you hear that a program, a process, or a procedure has been developed, a red flag should go up in your mind what about the implementation of that program, process, or procedure? The same goes for the principle of matching corrective actions with deficiencies. A good leader, almost without thinking, automatically looks for alignment of corrective actions to problems. I recall one large production facility that was having a recurring problem of their training instructors failing their requalification examinations. Their corrective actions for this were to change a procedure and develop an indicator. Think about that. Mary the instructor fails her requalification exam. You have made a change to a procedure and set up an indicator. What have you done to prevent Mary from failing her requalification exam again? I would argue, nothing. One characteristic I have found in organizations that are prone to develop programs and procedures is that they are often drowning in paperwork. Paper is generally useless in preventing problem occurrence, so a problem occurs, paper is developed, another problem occurs, and more paper is developed, ad nauseam.

Remember that the key takeaway here is to fix problems effectively.

# COACH, DON'T CHEERLEAD

*"Coaching is unlocking a person's potential to maximize their own performance."*

-Timothy Gallwey

A leader is a coach, not a cheerleader. What's the difference? In the definition of cheerleading, any dictionary will at some point include the words "cheering, chanting, and dancing." Now imagine you are in charge of a group of people who are not performing well and consider how your workforce behaviors might change if you start cheering and chanting and dancing. Enough said about cheerleading. Now on to coaching. In the opening quote of this section, Mr. Gallwey, an American author who has written several books on a methodology of coaching, captures the point of coaching remarkably well. Having said that, I might describe it using a few different words. I believe that coaching is the act of helping those who work for you to meet your expectations. My description matches that of Mr. Gallwey if I expect that you perform to your maximum potential, and that is what I expect. I would go on to suggest that any leader do this by first clearly stating and ensuring people know what your expectations are, as discussed earlier, and then helping them by providing feedback based on observation and other monitoring. For example, if you want your employees to follow procedures, and you say, "I expect you to follow procedures" you are wasting your time. The operative word in my advice is to CLEARLY state your expectations. For example, do you expect the workers to have a procedure they're using in hand at all times? Do you want them to use some technique to ensure they don't miss a step? The "circle/slash method is an effective technique used in many organizations where procedure adherence is important. When a step is started, circle the step number. When the step is complete, draw a slash through the circle. This reduces the likelihood of missing a step. If you expect all of this, and you should if procedure adherence is important, then you should state that. Show the employee how to do it. Coach them to perform the way you expect them to perform.

# PROVIDE INSIGHT

*"When you want wisdom and insight as badly as you want to breathe, it is then you shall have it."*

-Socrates

Effective leaders are valued for many things, but one of these things, in particular, is the value they add to reports, both verbal and written. This value often comes in the form of insight. The best leaders are often recognized for the valuable insight they provide, and their ability to provide it is proportional to the time they spend and have spent in the trenches. Having been with the workers allows one to see the kinds of problem aspects that lead to useful insight. For example, when reporting on a condition, such as a facility's cleanliness, to someone higher in your organization than you are, you can report the facts on the state, i.e., you can say Statement 1: "The facility is dirty", or you can add considerable value to your report by providing insight. Insight is defined by some sources as "seeing into a person or situation". It is valuable because by seeing into a situation, one can understand it better, including possibly the reasons for or contributors to the situation, and thus the insight is valuable in determining what corrective actions might need to be taken. For example, consider statement 2: "The facility is dirty because there is no ownership of the various areas. Some companies have assigned each key area to an individual or a group of individuals, thus promoting ownership of the areas." Which statement, 1 or 2, do you find more helpful? The opening quote by our good friend Socrates, which essentially puts insight and wisdom in the same basket, as it should be, is also a clever way of communicating the important point that anyone can get insight, but one has to really want it. He also, again cleverly, by comparing the desire for insight to the need to breathe, sets insight on a shelf of importance that is appropriately high.

It is easy to say one should include insight in reports, both verbal and written, but it is another thing to actually include the insight. It takes a sensitivity to the need for and value of insight, and it requires as well the effort needed to develop and include the insight. Again, back to our example of the dirty facility. If you were to identify this deficiency yourself, depending on the amount and type of experience you have, you might not have the insight to include in your report. This is not a shortfall, it's an opportunity to consider another valuable aspect of insight, and that is your own use of it. Get in the habit of looking for insight. Set a clear expectation that you want it in reports from others, but look for insight into issues even if you are not going to make a report on them. Maybe it's your responsibility to fix the issue. If you are in the habit of looking for insight, you are likely to ask more questions and probe into any issues you identify. Insight is valuable to everyone, including yourself.

# EFFECTIVELY USE THE CORPORATE FUNCTION.

*"Always remember that the corporate office serves an important function, but they PRODUCE nothing and our company's reason for existence is to PRODUCE."*

-WTS

The world of business practices is not unlike the fashion industry in that it is equally susceptible to fads, these being practices or interests, as described in Merriam Webster's, as being "popular for a short time". The numerous fads that have seeped into the business community include MBWA or management by walking around; flattening organizational structure; and use of buzz words such as "reaching out," "clickbait", and "game-changer". Another fad, and one I have experienced on multiple occasions, having worked with and evaluated numerous corporate organizations, deals with attempts to achieve a more cost-effective organization. Corporate Executive A says we can save a lot of money by combining certain functions currently performed at production sites, like administration, recruiting, regulatory services, and maybe even engineering, removing them from the place of production and having those functions performed for the facilities by staff at the corporate office. This "economy of scale" is intended to leverage resources and "do more with less" (more faddish buzzwords). Corporate Executive B, whose company currently has a large corporate function, says we can save a lot of money by doing the opposite, decentralizing, moving the above functions to the production sites, and significantly reducing the number of people needed for support. This, says Executive B, will also solve the problem of the remote corporate staff not having ownership of facility problems and/or not even knowing what those problems are, and being interested in those problems only if they occur during the 9 AM to 5 PM corporate working hours. Which of these is right? Maybe neither. An effective leader recognizes that one does not solve problems by reorganizing and that changing organizational structure because of production plant performance problems often produces nothing more than the impression that something is being done about the problems. The Nuclear Mustang

Leader recognizes as well that, as discussed at length in *The Observant Eye,* and earlier in this book, the only way to change performance is to change behaviors.

I have seen the transitions above, described as being carried out by Executives A and B, innumerable times. Except in those cases where the change was accompanied by other actions that caused behavior changes, a fingerless person could count the successes on one hand.

A real leader recognizes that a corporate office exists primarily to support the production element of the company. I once had a boss who, to make the above point in an electric power-producing company, said to the corporate staff, "The only thing corporate operates around here are doorknobs". The point of this statement was to emphasize that any unnecessary burden put on the operating units of the company by the corporate office has a negative effect on the company's bottom line and is to be avoided.

A real leader also recognizes that it is important for those on the production line to know that the corporate office does play an important role, and the more this is realized, the greater their empathy and appreciation, and the more effective the teamwork within the company. Take the time to educate the producers on what the corporate function is for and what they do, and the corporate staff on the key role that the producers play.

# NEVER HAVE A BAD DAY

*"The only difference between a good and bad day is your attitude. The choice is yours."*

-Dennis S. Brown

As those who have battled in the arena of warriors-in-charge know, there is never a shortage of opportunities to have a bad day. The more challenging the job, the greater the number of opportunities. Maybe the message of ensuing budget cuts or layoffs is the precipitating event. Maybe you didn't get enough sleep the night before, for whatever reason. Maybe you have a personal issue in your life that has sucked the happiness out of your heart. Maybe the job is just really hard and it seems at times that you can't get your head above water. Any of these might be true because sometimes that's just how life is. As one of my submarine mates once succinctly explained life to me as I was whining about having a bad day, "That's the way it is. "Life is a bitch and then you die." This crude but unarguably accurate description of life doesn't mean we have to drag everybody else down into depression with us. Not if we are real leaders.

Such leaders never have a bad day. The opening
quote of this section, attributed by some to the
motivational speaker and author Dennis S.
Brown, but conveyed in various forums as being
spoken by several other people earlier than when
used by Mr. Brown, is valuable nonetheless.
Recognize that attitude is only evident in displays
- by external signs, or verbal and nonverbal
signals, that one exhibits. So, if you're
downhearted, for whatever reason, no one knows
it unless you walk around displaying "Woe is
me" signs, in how you act, what you say, and
even in what you don't do. If you refuse to
display these signs, no one will ever know you're
downhearted, because, again, what you display is
how you are seen. Why is how you are seen
important? Because the attitudes of those in
charge are like viruses. They spread. They infect.
They're highly contagious. When you think
you've had a bad day and you interact with those
you hope to lead, suck it up. Project the upbeat
attitude you want them to have. Certainly, don't
saddle them with your problems, and don't even
remind them of problems that they might have. It
is commonly known that attitude affects a
worker's performance as well as the morale of
those around him. Generally, workers with good

attitudes have strong performance, and workers with poor attitudes do not.

Unfortunately, I have had numerous opportunities to see this phenomenon and its effect, usually when one of the submarines on which I served was leaving for a prolonged deployment. Depression is probably too strong a word, but it's close. Departing on deployments that likely will go for months at a time, without the ability to communicate with families and loved ones, to experience holidays, to put little ones to bed, to just spend time with one's spouse, this is the stuff of a bad day. Inside I could feel the need to get out one of those "Woe is me" signs and wave it around so everyone would know how I felt. But I didn't, because I knew that if I did, the submarine mates in my charge, who were on the verge of the same feelings, would resonate with my signals, and our performance would suffer. Submarining requires a hundred percent attention, attention to detail. Is that valve in the right position, that switch? Is the water in that tank enough? Too much? Is the reading on that gauge too high? There is no time for self-sympathy. Again, this falls under one of the previously discussed maxims. Don't stress over things that you cannot control. Be upbeat. Focus on the positive. Show some humor. And don't ever have a bad day.

# DEAL WELL WITH CRITICISM

*"Criticism is something we can avoid easily - by saying nothing, doing nothing, and being nothing."*

-Aristotle
(or Elbert
Hubbard?)

Give some thought to the following question: What is something from which everyone can benefit, some badly need, many ask for, and essentially no one really wants? No, this is not a riddle. It is, rather, an interesting phenomenon within the ranks of those who would be leaders. It's a valid question, and the answer will rarely be found anywhere but in this book, because, as the question states, even though people ask for it, they don't really want it. That "thing" is criticism. Criticism, although something we can all benefit from, is also something that we, by our human nature, try to avoid, sometimes even subconsciously. It can be avoided, but (as some people incorrectly think) Aristotle once said the quote leading off this section, "Criticism is something we can avoid easily - by saying nothing, doing nothing, and being nothing." These words were not mouthed by Aristotle but rather were written by Elbert Hubbard, an American writer, artist, and philosopher who lived in the early 1900's and was, in part, famous for writing the famous, "A Message to Garcia", an essay, which although not true, was highly effective in communicating the importance of initiative in carrying out a tough assignment. All that said, regardless of its source, the quote on criticism concisely communicates its importance.

This book is not about travel, but before continuing on the subject of criticism, a few words about going to a state that a leader should never enter, and should be able to quickly recognize should she be seen in it. That is the state of denial. I promised earlier that I would come back to the topic of denial, and here, in a discussion of providing critical feedback, is the place of that return. I cannot recall the number of denial-prone executives with whom I have interacted while giving them candid feedback following an assessment I performed on their organization, but I know that number is large, surprisingly large. The denial would often expose itself in words from the executive like, "We just didn't represent ourselves the way we should have". Believe me, representing oneself was usually not the problem. Below I will share my thoughts on what you should do when you believe criticism of you is wrong. Before reading that, a few words on making sure your assessment of "wrong" is not wrong. If upon receiving criticism that you think is wrong, your first reaction should be to candidly consider whether or not you might be in that, to-be-avoided-at-all-costs, state of denial. The criticism might be valid but you don't want to accept that possibility because you are hurt by the criticism. You want to believe you are better than that. You want to believe that the criticism is wrong. Having this feeling is a valuable experience because it will help you to understand the thinking and feelings of others to whom you provide criticism. Learn from this feeling and move on to deal with the criticism.

The best leaders thrive on criticism because they know that dealing with it will make them better, even if the criticism is wrong. If what they are told is wrong in their minds, they still take action because they realize that they have done something to lead someone to formulate a critical impression of them, and that is what they will fix. They can't get enough of criticism because they are striving for excellence in everything they do, and criticism is what fuels them to move on. They are not bothered by criticism because whatever is said will be less critical than what they receive from their most judgmental critic themselves. They are never satisfied with what they have done because they know they can always do better. They never achieve excellence in their performance, but by striving for it, they get better and better, and better.

The header on this section doesn't say, "accept criticism well", it states "deal well with criticism". Dealing with it means accepting it, but also giving it, to those below her, above her, or next to her, her colleagues. A few words on each. And remember, at the outset of this section, we said that no one wants criticism. When providing criticism to subordinates, an effective leader takes into consideration the experience of the receiver. A direct mortar blast of critical feedback to a new employee can be devastating and motivation stifling because, if that is among the first feedback he receives, it can be seen as a picture of that person's performance in total. The receiver probably doesn't have the experience that the leader in this discussion has and can rely on to know that he or she is not a total failure. On the other hand, if the receiver does have considerable experience, or has been given this criticism on more than one occasion in the past, then the criticism may need to be sharpened even further to be more effective. If this stronger approach doesn't work, the leader should be looking for a solution to what is a different problem. Maybe the person doesn't have the capability, is in the wrong position, or has an intractable attitude. The solutions to those kinds of problems are the stuff of a different section of this book.

How do you provide criticism to your peers? You don't. Remember, no one wants criticism. So even if it's valid, the cost in feelings and relationships far surpasses any individual gain for that person from the criticism. And why would you think you have the authority or professional expertise to give criticism anyway? Keep your thoughts to yourself except to complement your peer on whatever you thought was done well.

And finally, how do you give criticism to your boss? Simply put, again, you do not, regardless of what the boss says she wants. When trying to lead the boss, see the earlier section on leading those below, beside, and above you.

# DON'T TOLERATE HAPPY TALK

*"Happy Talk eventually leads to a state of unhappiness - about work, finances, and performance in general."*

-WTS

The meeting was called to order as those in attendance were still exchanging pleasantries. Everyone in the room appeared to be comfortable as the status of the organization was reported, and all was reported as going well. No major problems were either in existence or foreseen to be in the near future. A few less-than-serious problems were mentioned with the caveat that essentially nothing could be done about these. Reported results stemming from the activities that were being discussed, although not particularly good, were presented as being the best that could be achieved given the resources available. There were no dissenting opinions. Everyone seemed to agree with the actions taken or planned. A warm sense of satisfaction permeated the room, and all seemed somewhat anxious to end the meeting and get on with other business.

The above is a good example of Happy Talk, the kind of talk that goes on in gatherings when those in charge have little idea of what's going on in their organizations, and are, therefore, not fully in touch with reality.

Don't bother looking up the definition of Happy Talk. You won't find it. You will find some terms that seem similar, like status-quo bias, and optimism bias. But those terms are overly formal and a bit off-target. If you want to hear about Happy Talk, spend some time on one of those vessels that typically travel hundreds of feet beneath the surface of the ocean, and are propelled by nuclear power. At least that was the case during my time "on the boats." It was quickly identified as such by the crew and to say it was discouraged would be an understatement. Happy talk was something that under no circumstances would be tolerated, by just about anyone. Don't get me wrong. Happy Talk does have positives. It gives one a feeling of comfort. It contributes to an atmosphere in which nothing more needs to be done. No additional effort is needed. Things are just fine; as good as they can be. The problem with Happy Talk is that it stifles advancement, improvement, and effort to do better than we have done in the past. That was why boat sailors didn't tolerate it.

Recognize that when I say, a good leader never tolerates Happy Talk, I am not saying don't be happy. I'm saying that if truthfulness and openness are core principles of an organization, as they should be, then problems and challenges will be part of any discussions because problems and challenges are part of life if the expectation is to achieve excellence in all aspects of the business. That's a high bar. Working to achieve it can and should be very satisfying, but problems will occur. It is how these problems are recognized and addressed that differentiates the best from the less-than-best. It is an effective leader that establishes the atmosphere in which we continuously seek to improve performance in everything we do.

# BEWARE OF THE HAZARDS OF A CLAY LAYER

*"Having a clay layer in management is a solvable problem. The real issue is having one and not knowing it."*

-WTS

The term "clay layer" has been around in management books for some time, its popularity slowly giving way to some of the more faddish new ideas in management theory. But it is a concept, with which experienced leaders are familiar and one well worth being aware of and able to deal with.

For those of a non-agricultural bent, clay is one of about five different types of soil. Although it is high in nutrients, it can be a killer for flora of all types if it arranges itself in a particular configuration. That configuration, much like the one applied to management organizations, is a compact layer anywhere from a few inches to more than a foot in depth. Since clay is made of the smallest particles in the various types of soil, the soil is very dense, bonds together, and lends itself well to compacting and providing an impenetrable barrier that precludes the flow of water to the roots of plants, decreases the needed nitrogen in plants, and hinders root penetration of those plants attempting to grow above the clay layer.

With that explanation in mind, picture an organization chart in place of a cross-section of soil. Mentally sketch some parallel lines across the chart, placing them somewhere in the middle of the vertical organizational spread. Instead of plant roots being hindered by the clay, visualize communication, both up and down the organization as being the counterpart of plant roots. Such is the clay layer in the soil of management.

I once headed a technical organization of just under 1,000 permanent employees. On occasion, special evolutions could swell these ranks up into the several thousand range because of the addition of contractors.

Shortly after I assumed the position, I realized that neither the messages from me nor those from my direct reports were getting to the workers. I subsequently found there was a dense clay layer that stretched over the supervisor level and blocked the nutrients and water of good communication from getting from me to the lowest working level. It had built up over the years because of a lack of trust in the frequently changing management. It was apparent to me that the workers wanted to keep their jobs, so they operated the facility as best they could. The supervisors, lacking support from management and being generally weak in supervisory skills, commiserated with the workers to the point that the professional separation between worker and supervisor was heavily blurred. The root of (or insight into) the problem was a complete lack of trust in management. Regular turnover of management exacerbated the problem, and this turnover resulted from an inability of the managers to effectively operate the facility with what was clearly an inadequate budget. As a result of this money starvation, the workers had learned to live with the hand they had been dealt. Supervisors and workers had banded together as a survival mechanism. Management was not a significant factor in their lives. The workers knew they didn't have to do what their manager said, because before long, he or she would be gone. Management in general referred to the workers as the "wee bees", because of their position that "we be here when you (the managers) be gone". The underfunding of the facility led to the workers and their supervisors learning to keep the facility operating, even when a more prudent move would be to shut it down and get various problems properly addressed. This had happened so frequently over time, that it became the accepted way of doing business, the culture, so to speak. To further support this approach, the workers learned to ignore problems. The standard response to any comment about one problem or another typically began with, "That's not a problem because …" The owners of the company had no idea what was going on in the trenches because they had never gone there. (This point on being in the trenches keeps coming up.) They were satisfied because the facility was operating and they didn't have to expend the money it would have taken to resolve the problems that came up on an almost daily basis. Of course, this approach can only work for so long. Eventually, the growing problems overtake the organization. That was happening about the time I arrived. I only share this story to illustrate the issues of a clay layer, to caution the aspiring leader to be aware of the potential for such a layer to form and to give some understanding of the kind of consequences that can result from such a layer.

Breaking through such a layer is not easy, and it takes time. The essential tools for eliminating it consist of first of all becoming aware that the layer exists. This can only be accomplished by managers being Mustangs, spending time in the field, assessing performance, listening to the problems of the workers, and checking the effectiveness of downward communication. At the same time, these managers will be building worker trust in leadership by demonstrating management support for the workforce.

# RUN TOWARD PROBLEMS

*"If you see a problem, own it"*

-WTS

This final piece of advice brings us back to the story I told in the opening pages of this book, about the submariner who jumped into the flooding compartment, shutting and sealing the door behind him. I talked in later pages about why he did that. One way to look at this, for someone wondering what a submarine hero has to do with a corporate manager, is that the flooding in a submarine is a "problem". The submariner, by his action, clearly demonstrated that he owned the problem. He ran toward it. He didn't wait for someone else to fix it. At the most basic level, he was motivated by the fact that if the submarine went to crush depth while he was waiting for someone else to fix the problem, everyone would die, including him. But he didn't even have to think about these consequences. The actions he would take had been inculcated in him through extensive training and by the examples set for him by others on his crew. He acted automatically. The concept of every single member of a team either working together or dying together had been deeply ingrained in his training, his exercises, his drills, his discussions with and observations of his submarine-mates. This thinking was just natural for him. Your office might not be capable of going down to a depth where everybody dies in the crush of hundreds of pounds per square inch of sea pressure. It might not wait for someone else to fix it. That seemingly suicidal submariner set an example of really owning a problem and putting the boat, or the company first. When a problem occurs in your company, do you feel that kind of ownership? Are you motivated to run toward it, instead of waiting for someone else to fix it? You are if you're a good leader, especially if you believe in and foster Nuclear Mustang Leadership. This is just another way of saying, act like you own the company. If you're going to spend money, it's not company money, it's your money. Would you spend the money if it was yours? Nurture this kind of thinking in your day-to-day activities and you'll find it becoming a habit to think and act this way without even realizing it. When you do, you will be a leader and fostering Nuclear Mustang Leadership.

# CLOSING THOUGHTS

In the previous pages, I have attempted to provide what are in my mind the most important recommendations for achieving a type of leadership that I know works and works well. Behaviors aligned with these recommendations have moved me from a young man with minimal education, no skills, and a net worth of minus 115 dollars to a highly successful executive with an advanced degree in Electrical Engineering and advanced management education at the Harvard Business School, having been the Chief Engineer on a nuclear-powered submarine, a vice president and company officer in a nuclear power business, and a highly paid management consultant, now retired for the third time.

If you think adhering to the principles outlined in this book is hard, you are correct. If you think the pursuit of this is too hard for you, I encourage you to pursue a career path that doesn't require you to play a leadership role. People who think anything is too hard do not have the makeup to be real leaders.

But if you choose to be among the best of leaders, then read this book again. Exert every effort to follow the advice provided. Continually seek to improve your performance. And most importantly, never give up.